John Ruskin

Proserpina

Studies of wayside flowers, while the air was yet pure among the Alps. Vil. 1, Third Edition

John Ruskin

Proserpina
Studies of wayside flowers, while the air was yet pure among the Alps. Vil. 1, Third Edition

ISBN/EAN: 9783337111601

Printed in Europe, USA, Canada, Australia, Japan

Cover: Foto ©berggeist007 / pixelio.de

More available books at **www.hansebooks.com**

Blossoming — and stricken in days

PROSERPINA.

STUDIES OF WAYSIDE FLOWERS,

WHILE THE AIR WAS YET PURE

AMONG THE ALPS, AND IN THE SCOTLAND AND ENGLAND WHICH MY FATHER KNEW.

BY

JOHN RUSKIN, LL.D.,

HONORARY STUDENT OF CHRIST CHURCH, AND SLADE PROFESSOR OF FINE ART.

"Oh—Prosérpina!
For the flowers now, which frighted, thou let'st fall
From Dis's waggon."

VOLUME I.

THIRD EDITION.

GEORGE ALLEN,
ORPINGTON AND LONDON.
1897.

CONTENTS OF VOL. I.

CHAPTER VIII.

CHAPTER IX.

CHAPTER X.

CHAPTER XI.

CHAPTER XII.

CHAPTER XIII.

CHAPTER XIV.

INDEX I.

INDEX II.

INDEX III.

LIST OF ILLUSTRATIONS.

VOL. I.

PROSERPINA.

INTRODUCTION.

BRANTWOOD, 14*th March*, 1874.

YESTERDAY evening I was looking over the first book in which I studied Botany,— Curtis's Magazine, published in 1795 at No. 3, St. George's Crescent, Blackfriars Road, and sold by the principal booksellers in Great Britain and Ireland. Its plates are excellent, so that I am always glad to find in it the picture of a flower I know. And I came yesterday upon what I suppose to be a variety of a favourite flower of mine, called, in Curtis, "the St. Bruno's Lily."

I am obliged to say "what I suppose to be a variety," because my pet lily is branched,* while this is drawn as unbranched, and especially stated to be so. And the page of text, in which this

* At least, it throws off its flowers on each side in a bewilderingly pretty way; a real lily can't branch, I believe: but, if not, what is the use of the botanical books saying "on an unbranched stem"?

statement is made, is so characteristic of botanical books, and botanical science, not to say all science as hitherto taught for the blessing of mankind; and of the difficulties thereby accompanying its communication, that I extract the page entire, printing it, opposite, as nearly as possible in facsimile.

Now you observe, in this instructive page, that you have in the first place, eight names given you for one flower; and that, among these eight names, you are not even at liberty to make your choice, because the united authority of Haller and Miller may be considered as an accurate balance to the single authority of Linnæus; and you ought therefore for the present to remain, yourself, balanced between the sides. You may be farther embarrassed by finding that the Anthericum of Savoy is only described as growing in Switzerland. And farther still, by finding that Mr. Miller describes two varieties of it, which differ only in size, while you are left to conjecture whether the one here figured is the larger or smaller; and how great the difference is.

Farther, If you wish to know anything of the habits of the plant, as well as its eight names, you are informed that it grows both at the bottoms of the mountains, and the tops; and that, with us, it

ANTHERICUM LILIASTRUM. SAVOY ANTHERICUM, OF ST. BRUNO'S LILY.

Claſs and Order.

HEXANDRIA MONOGYNIA.

Generic Character.

Cor. 6-petala, patens. *Capſ.* ovata.

Specific Character and Synonyms.

ANTHERICUM *Liliaſtrum* foliis planis, ſcapo ſimpliciſſimo, corollis campanulatis, ſtaminiſibus declinatis. *Linn. Syſt. Vegetab. ed.* 14. *Murr. p.* 330. *Ait. Kew. v.* 1. *p.* 449.

HEMEROCALLIS floribus patulis ſecundis. *Hall. Hiſt. n.* 1230.

PHALANGIUM magno flore. *Bauh. Pin.* 29.

PHALANGIUM Allobrogicum majus. *Cluſ. cur. app. alt.*

PHALANGIUM Allobrogicum. The Savoye Spider-wort. *Park. Parad. p.* 150. *tab.* 151. *f.* 1.

Botaniſts are divided in their opinions reſpecting the genus of this plant; LINNÆUS conſiders it as an *Anthericum*, HALLER and MILLER make it an *Hemerocallis*.

It is a native of Switzerland, where, HALLER informs us, it grows abundantly in the Alpine meadows, and even on the ſummits of the mountains; with us it flowers in May and June.

It is a plant of great elegance, producing on an unbranched ſtem about a foot and a half high, numerous flowers of a delicate white colour, much ſmaller, but reſembling in form thoſe of the common white lily, poſſeſſing a conſiderable degree of fragrance, their beauty is heightened by the rich orange colour of their antheræ; unfortunately they are but of ſhort duration.

MILLER deſcribes two varieties of it differing merely in ſize.

A loamy ſoil, a ſituation moderately moiſt, with an eaſtern or weſtern expoſure, ſuits this plant beſt; ſo ſituated, it will increaſe by its roots, though not very faſt, and by parting of theſe in the autumn, it is uſually propagated.

PARKINSON deſcribes and figures it in his *Parad. Terreſt.*, obſerving, that "divers allured by the beauty of its flowers, had "brought it into theſe parts."

flowers in May and June,—but you are not told when, in its native country.

The four lines of the last clause but one, may indeed be useful to gardeners; but—although I know my good father and mother did the best they could for me in buying this beautiful book; and though the admirable plates of it did their work, and taught me much, I cannot wonder that neither my infantine nor boyish mind was irresistibly attracted by the text, of which this page is one of the most favourable specimens; nor, in consequence, that my botanical studies were—when I had attained the age of fifty—no farther advanced than the reader will find them in the opening chapter of this book.

Which said book was therefore undertaken, to put, if it might be, some elements of the science of botany into a form more tenable by ordinary human and childish faculties; or—for I can scarcely say I have yet any tenure of it myself—to make the paths of approach to it more pleasant. In fact, I only know, of it, the pleasant distant effects, which it bears to simple eyes; and some pretty mists and mysteries, which I invite my young readers to pierce, as they may, for themselves,—my power of guiding them being only for a little way.

Pretty mysteries, I say, as opposed to the vulgar and ugly mysteries of the so-called science of botany,—exemplified sufficiently in this chosen page. Respecting which, please observe farther:— Nobody—I can say this very boldly—loves Latin more dearly than I; but, precisely because I do love it (as well as for other reasons), I have always insisted that books, whether scientific or not, ought to be written either in Latin, or English; and not in a doggish mixture of the refuse of both.

Linnæus wrote a noble book of universal Natural History in Latin. It is one of the permanent classical treasures of the world. And if any scientific man thinks his labours are worth the world's attention, let him, also, write what he has to say in Latin, finishedly and exquisitely, if it take him a month to a page.*

But if—which, unless he be one chosen of millions, is assuredly the fact—his lucubrations are only of local and temporary consequence, let him write, as clearly as he can, in his native language.

This book, accordingly, I have written in English; (not, by the way, that I *could* have written it in anything else—so there are small thanks to me;)

* I have by happy chance just added to my Oxford library the poet Gray's copy of Linnæus, with its exquisitely written Latin notes, exemplary alike to scholar and naturalist.

and one of its purposes is to interpret, for young English readers, the necessary European Latin or Greek names of flowers, and to make them vivid and vital to their understandings. But two great difficulties occur in doing this. The first, that there are generally from three or four, up to two dozen, Latin names current for every flower; and every new botanist thinks his eminence only to be properly asserted by adding another.

The second, and a much more serious one, is of the Devil's own contriving—(and remember I am always quite serious when I speak of the Devil,)—namely, that the most current and authoritative names are apt to be founded on some unclean or debasing association, so that to interpret them is to defile the reader's mind. I will give no instance; too many will at once occur to any learned reader, and the unlearned I need not vex with so much as one: but, in such cases, since I could only take refuge in the untranslated word by leaving other Greek or Latin words also untranslated, and the nomenclature still entirely senseless,—and I do not choose to do this,—there is only one other course open to me, namely, to substitute boldly, to my own pupils, other generic names for the plants thus faultfully hitherto titled.

As I do not do this for my own pride, but

honestly for my readers' service, I neither question nor care how far the emendations I propose may be now or hereafter adopted. I shall not even name the cases in which they have been made for the serious reason above specified; but even shall mask those which there was real occasion to alter, by sometimes giving new names in cases where there was no necessity for such a kind. Doubtless I shall be accused of doing myself what I violently blame in others. I do so; but with a different motive—of which let the reader judge as he is disposed. The practical result will be that the children who learn botany on the system adopted in this book will know the useful and beautiful names of plants hitherto given, in all languages; the useless and ugly ones they will not know. And they will have to learn one Latin name for each plant, which, when differing from the common one, I trust may yet by some scientific persons be accepted, and with ultimate advantage.

The learning of the one Latin name—as, for instance, Gramen striatum—I hope will be accurately enforced always;—but not less carefully the learning of the pretty English one—"Ladie-lace Grass"—with due observance that "Ladies' laces hath leaves like unto Millet in fashion, with many white vaines or ribs, and silver strakes running along through the middest of the leaves, fashioning the same like to laces of white and green silk, very beautiful and faire to behold."

I have said elsewhere, and can scarcely repeat too often, that a day will come when men of science will think their names disgraced, instead of honoured, by being used to barbarise nomenclature; I hope therefore that my own name may be kept well out of the way; but, having been privileged to found the School of Art in the University of Oxford, I think that I am justified in requesting any scientific writers who may look kindly upon this book, to add such of the names suggested in it as they think deserving of acceptance, to their own lists of synonyms, under the head of "Schol. Art. Oxon."

The difficulties thrown in the way of any quiet private student by existing nomenclature may be best illustrated by my simply stating what happens to myself in endeavouring to use the page above facsimile'd. Not knowing how far St. Bruno's Lily might be connected with my own pet one, and not having any sufficient book on Swiss botany, I take down Loudon's Encyclopædia of Plants, (a most useful book, as far as any book in the present state of the science *can* be useful,) and find, under the head of Anthericum, the Savoy Lily indeed, but only the following general information:—"809. Anthericum. A name applied by the Greeks to the stem of the asphodel, and not misapplied to this set of plants, which in some sort resemble the asphodel. Plants with fleshy leaves, and spikes of bright *yellow* flowers, easily cultivated if kept dry."

Hunting further, I find again my Savoy Lily called a spider-plant, under the article Hemerocallis, and the only information which the book gives me under Hemerocallis, is that it means 'beautiful day' lily; and then, "This is an ornamental genus of the easiest culture. The species are remarkable among border flowers for their fine *orange*, *yellow*, or *blue* flowers. The Hemerocallis cœrulea has been considered a distinct genus by Mr. Salisbury, and called Saussurea." As I correct this sheet for press, however, I find that the Hemerocallis is now to be called 'Funkia,' "in honour of Mr. Funk, a Prussian apothecary."

All this while, meantime, I have a suspicion that my pet Savoy Lily is not, in existing classification, an Anthericum, nor a Hemerocallis, but a Lilium. It is, in fact, simply a Turk's cap which doesn't curl up. But on trying 'Lilium' in Loudon, I find no mention whatever of any wild branched white lily.

I then try the next word in my specimen page of Curtis; but there is no 'Phalangium' at all in Loudon's index. And now I have neither time nor mind for more search, but will give, in due place, such account as I can of my own dwarf branched lily, which I shall call St. Bruno's, as well as this Liliastrum—no offence to the saint, I hope. For it grows very gloriously on the limestones of Savoy, presumably, therefore, at the Grande Chartreuse; though I did not notice it there, and made a very unmonkish use of it when I gathered it last: —There was a pretty young English lady at the table-d'hôte, in the Hotel du Mont Blanc at St. Martin's,* and I wanted to get speech of her, and didn't know how. So all I could think of was to go half-way up the Aiguille de Varens, to gather St. Bruno's lilies; and I made a great cluster of them, and put wild roses all round them as I came down. I never saw anything so lovely; and I thought to present this to her before dinner,—but when I got down, she had gone away to Chamouni. My Fors always treated me like that, in affairs of the heart.

I had begun my studies of Alpine botany just eighteen years before, in 1842, by making a careful drawing of wood-sorrel at Chamouni; and

* It was in the year 1860, in June.

Line-Study I

ERICA TETRALIX.

bitterly sorry I am, now, that the work was interrupted. For I drew, then, very delicately; and should have made a pretty book if I could have got peace. Even yet, I can manage my point a little, and would far rather be making outlines of flowers than writing; and I meant to have drawn every English and Scottish wild flower, like this cluster of bog heather opposite,*—back, and profile, and front. But 'Blackwood's Magazine,' with its insults to Turner, dragged me into controversy; and I have not had, properly speaking, a day's peace since; so that in 1868 my botanical studies were advanced only as far as the reader will see in next chapter; and now, in 1874, must end altogether, I suppose, heavier thoughts and work coming fast on me. So that, finding among my note-books, two or three, full of broken materials for the proposed work on flowers; and, thinking they may be useful even as fragments, I am going to publish them in their present state,—only let the reader note that while my other books endeavour, and claim, so far as they reach, to give trustworthy knowledge of their subjects, this one only shows how such knowledge may

* Admirably engraved by Mr. Burgess, from my pen drawing, now at Oxford. By comparing it with the plate of the same flower in Sowerby's work, the student will at once see the difference between attentive drawing, which gives the cadence and relation of masses in a group, and the mere copying of each flower in an unconsidered huddle.

be obtained; and it is little more than a history of efforts and plans,—but of both, I believe, made in right methods.

One part of the book, however, will, I think, be found of permanent value. Mr. Burgess has engraved on wood, in reduced size, with consummate skill, some of the excellent old drawings in the Flora Danica, and has interpreted, and facsimile'd, some of his own and my drawings from nature, with a vigour and precision unsurpassed in woodcut illustration, which render these outlines the best exercises in black and white I have yet been able to prepare for my drawing pupils. The larger engravings by Mr. Allen may also be used with advantage as copies for drawings with pen or sepia.

ROME, 10*th May* (*my father's birthday*).

I found the loveliest blue asphodel I ever saw in my life, yesterday, in the fields beyond Monte Mario, —a spire two feet high, of more than two hundred stars, the stalks of them all deep blue, as well as the flowers. Heaven send all honest people the gathering of the like, in Elysian fields, some day!

CHAPTER I.

MOSS.

DENMARK HILL, 3rd November, 1868.

1. IT is mortifying enough to write,—but I think thus much ought to be written,—concerning myself, as 'the author of Modern Painters.' In three months I shall be fifty years old: and I don't at this hour—ten o'clock in the morning of the two hundred and sixty-eighth day of my forty-ninth year—know what 'moss' is.

There is nothing I have more *intended* to know—some day or other. But the moss 'would always be there'; and then it was so beautiful, and so difficult to examine, that one could only do it in some quite separated time of happy leisure—which came not. I never was like to have less leisure than now, but I *will* know what moss is, if possible, forthwith.

2. To that end I read preparatorily yesterday, what account I could find of it in all the botanical books in the house. Out of them all, I get this

general notion of a moss,—that it has a fine fibrous root,—a stem surrounded with spirally set leaves, —and produces its fruit in a small case, under a cap. I fasten especially, however, on a sentence of Louis Figuier's, about the particular species, Hypnum:—

"These mosses, which often form little islets of verdure at the feet of poplars and willows, are robust vegetable organisms, which do not decay."*

3. "Qui ne pourrissent point." What do they do with themselves, then?—it immediately occurs to me to ask. And, secondly,—If this immortality belongs to the Hypnum only?

It certainly does not, by any means: but, however modified or limited, this immortality is the first thing we ought to take note of in the mosses. They are, in some degree, what the 'everlasting' is in flowers. Those minute green leaves of theirs do not decay, nor fall.

But how do they die, or how stop growing, then?—it is the first thing I want to know about them. And from all the books in the house, I can't as yet find out this. Meanwhile I will look at the leaves themselves.

4. Going out to the garden, I bring in a bit of

* "Histoire des Plantes." Ed. 1865, p. 416.

old brick, emerald green on its rugged surface, and a thick piece of mossy turf.

First, for the old brick: To think of the quantity of pleasure one has had in one's life from that emerald green velvet,—and yet that for the first time to-day I am verily going to look at it! Doing so, through a pocket lens of no great power, I find the velvet to be composed of small star-like groups of smooth, strong, oval leaves,—intensely green, and much like the young leaves of any other plant, except in this;—they all have a long brown spike, like a sting, at their ends.

5. Fastening on that, I take the Flora Danica,* and look through its plates of mosses, for their leaves only; and I find, first, that this spike, or strong central rib, is characteristic;—secondly, that the said leaves are apt to be not only spiked, but serrated, and otherwise angry-looking at the points;—thirdly that they have a tendency to fold together in the centre (Fig. 1 †); and at last, after an hour's work at them, it strikes me suddenly that

FIG. 1.

* Properly, Floræ Danicæ, but it is so tiresome to print the diphthongs that I shall always call it thus. It is a folio series, exquisitely begun, a hundred years ago; and not yet finished.

† Magnified about seven times. See note at end of this chapter.

they are more like pineapple leaves than anything else.

And it occurs to me, very unpleasantly, at the same time, that I don't know what a pineapple is!

Stopping to ascertain that, I am told that a pineapple belongs to the 'Bromeliaceæ'—(can't stop to find out what that means)—nay, that of these plants "the pineapple is the representative" (Loudon); "their habit is acid, their leaves rigid, and toothed with spines, their bracteas often coloured with scarlet, and their flowers either white or blue"—(what are their flowers like?) But the two sentences that most interest me, are, that in the damp forests of Carolina, the Tillandsia, which is an 'epiphyte' (*i.e.*, a plant growing on other plants), "forms dense festoons among the branches of the trees, vegetating among the black mould that collects upon the bark of trees in hot damp countries; other species are inhabitants of deep and gloomy forests, and others form, with their spring leaves, an impenetrable herbage in the Pampas of Brazil." So they really seem to be a kind of moss, on a vast scale.

6. Next, I find in Gray,* Bromeliaceæ, and—the very thing I want—"Tillandsia, the black *moss*, or long moss, which, *like most Bromelias*, grows on

* American,—'System of Botany,' the best technical book I have.

the branches of trees." So the pineapple is really a moss; only it is a moss that flowers but 'imperfectly.' "The fine fruit is caused by the consolidation of the imperfect flowers." (I wish we could consolidate some imperfect English moss-flowers into little pineapples then,—though they were only as big as filberts.) But we cannot follow that farther now; nor consider when a flower is perfect, and when it is not, or we should get into morals, and I don't know where else; we will go back to the moss I have gathered, for I begin to see my way, a little, to understanding it.

7. The second piece I have on the table is a cluster—an inch or two deep—of the moss that grows everywhere, and that the birds use for nest-building, and we for packing, and the like. It is dry, since yesterday, and its fibres define themselves against the dark ground in warm green, touched with a glittering light. Note that burnished lustre of the minute leaves; they are necessarily always relieved against dark hollows, and this lustre makes them much clearer and brighter than if they were of dead green. In that lustre—and it is characteristic of them—they differ wholly from the dead, aloe-like texture of the pineapple leaf; and remind me, as I look at them closely, a little of some conditions of

chaff, as on heads of wheat after being threshed. I will hunt down that clue presently; meantime there is something else to be noticed on the old brick.

8. Out of its emerald green cushions of minute leaves, there rise, here and there, thin red threads, each with a little brown cap, or something like a cap, at the top of it. These red threads shooting up out of the green tufts, are, I believe, the fructification of the moss; fringing its surface in the woods, and on the rocks, with the small forests of brown stems, each carrying its pointed cap or crest—of infinitely varied 'mode,' as we shall see presently; and, which is one of their most blessed functions, carrying high the dew in the morning; every spear balancing its own crystal globe.

9. And now, with my own broken memories of moss, and this unbroken, though unfinished, gift of the noble labour of other people, the Flora Danica, I can generalize the idea of the precious little plant, for myself, and for the reader.

All mosses, I believe, (with such exceptions and collateral groups as we may afterwards discover, but they are not many,) that is to say, some thousands of species, are, in their strength of existence, composed of fibres surrounded by clusters of dry *spinous* leaves, set close to the fibre they grow on. Out of this leafy stem descends a fibrous root, and ascends, in its season, a capped seed.

We must get this very clearly into our heads. Fig. 2, A, is a little tuft of a common wood moss of Norway,* in its fruit season, of its real size; but at present I want to look at the central fibre and its leaves accurately, and understand that first.

10. Pulling it to pieces, we find it composed of seven little company-keeping fibres, each of which, by itself, appears as in Fig. 2, B: but as in this, its real size, it is too small, not indeed for our respect, but for our comprehension, we magnify it, Fig. 2, C, and thereupon perceive it to be indeed composed of, *a*, the small fibrous root which sustains the plant; *b*, the leaf-surrounded stem which is the actual being, and main creature, moss; and, *c*, the aspirant pillar, and cap, of its fructification.

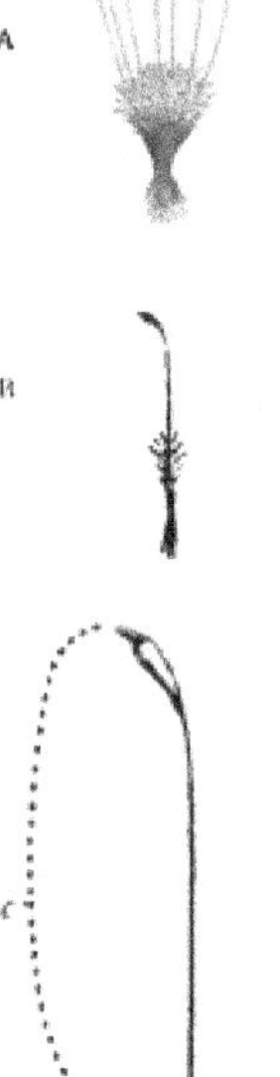

FIG. 2.

* 'Dicranum cerviculatum,' sequel to Flora Danica, Tab. MMCCX.

11. But there is one minor division yet. You see I have drawn the central part of the moss plant (*b*, Fig. 2,) half in outline and half in black; and that, similarly, in the upper group, which is too small to show the real roots, the base of the cluster is black. And you remember, I doubt not, how often, in gathering what most invited gathering, of deep green, starry, perfectly soft and living wood-moss, you found it fall asunder in your hand into multitudes of separate threads, each with its bright green crest, and long root of blackness.

That blackness at the root—though only so notable in this wood-moss and collateral species, is indeed a general character of the mosses, with rare exceptions. It is their funeral blackness;—that, I perceive, is the way the moss leaves die. They do not fall—they do not visibly decay. But they decay *in*visibly, in continual secession, beneath the ascending crest. They rise to form that crest, all green and bright, and take the light and air from those out of which they grew;—and those, their ancestors, darken and die slowly, and at last become a mass of mouldering ground. In fact, as I perceive farther, their final duty is so to die. The main work of other leaves is in their life,—but these have to form the earth out of which all other leaves are to grow. Not to cover the rocks with golden velvet only, but to fill their crannies with the dark earth, through which nobler creatures shall one day seek their being.

12. "Grant but as many sorts of mind as moss." Pope could not have known the hundredth part of the number of 'sorts' of moss there are; and I suppose he only chose the word because it was a monosyllable beginning with m, and the best English general expression for despised and minute structures of plants. But a fate rules the words of wise men, which makes their words truer, and worth more, than the men themselves know. No other plants have so endless variety on so similar a structure as the mosses; and none teach so well the Humility of Death. As for the death of our bodies, we have learned, wisely, or unwisely, to look the fact of that in the face. But none of us, I think, yet care to look the fact of the death of our minds in the face. I do not mean death of our souls, but of our mental work. So far as it is good *art*, indeed, and done in realistic form, it may perhaps not die; but so far as it was only good *thought*—good, for its time, and apparently a great achievement therein—that good, useful thought may yet in the future become a foolish thought, and then die quite away,—it, and the memory of

it,—when better thought and knowledge come. But the better thought could not have come if the weaker thought had not come first, and died in sustaining the better. If we think honestly, our thoughts will not only live usefully, but even perish usefully—like the moss—and become dark, not without due service. But if we think dishonestly, or malignantly, our thoughts will die like evil fungi,—dripping corrupt dew.

13. But farther. If you have walked moorlands enough to know the look of them, you know well those flat spaces or causeways of bright green or golden ground between the heathy rock masses; which signify winding pools and inlets of stagnant water caught among the rocks;—pools which the deep moss that covers them—*blanched*, not black, at the root,—is slowly filling and making firm; whence generally the unsafe ground in the moorland gets known by being *mossy* instead of heathy; and is at last called by its riders, briefly, 'the Moss': and as it is mainly at these same mossy places that the riding is difficult, and brings out the gifts of horse and rider, and discomfits all followers not similarly gifted, the skilled crosser of them got his name, naturally, of 'moss-rider,' or moss-trooper. In which manner the moss of Norway and Scotland has been a

taskmaster and Maker of Soldiers, as yet, the strongest known among natural powers. The lightning may kill a man, or cast down a tower, but these little tender leaves of moss—they and their progenitors—have trained the Northern Armies.

14. So much for the human meaning of that decay of the leaves. Now to go back to the little creatures themselves. It seems that the upper part of the moss fibre is especially *un*decaying among leaves; and the lower part, especially decaying. That, in fact, a plant of moss-fibre is a kind of persistent state of what is, in other plants, annual. Watch the year's growth of any luxuriant flower. First it comes out of the ground all fresh and bright; then, as the higher leaves and branches shoot up, those first leaves near the ground get brown, sickly, earthy,—remain for ever degraded in the dust, and under the dashed slime in rain, staining, and grieving, and loading them with obloquy of envious earth, half-killing them,—only life enough left in them to hold on the stem, and to be guardians of the rest of the plant from all they suffer;—while, above them, the happier leaves, for whom they are thus oppressed, bend freely to the sunshine, and drink the rain pure.

The moss strengthens on a diminished scale, intensifies, and makes perpetual, these two states,

—bright leaves above that never wither, leaves beneath, that exist only to wither.

15. I have hitherto spoken only of the fading moss as it is needed for change into earth. But I am not sure whether a yet more important office, in its days of age, be not its use as a colour.

We are all thankful enough—as far as we ever are so—for green moss, and yellow moss. But we are never enough grateful for black moss. The golden would be nothing without it, nor even the grey.

It is true that there are black lichens enough, and brown ones: nevertheless, the chief use of lichens is for silver and gold colour on rocks; and it is the dead moss which gives the leopard-like touches of black. And yet here again—as to a thing I have been looking at and painting all my life—I am brought to pause, the moment I think of it carefully. The black moss which gives the precious Velasquez touches, lies, much of it, flat on the rocks; radiating from its centres—powdering in the fingers, if one breaks it off, like dry tea. Is it a black species? or a black-parched state of other species, perishing for the sake of Velasquez effects, instead of accumulation of earth? and, if so, does it die of drought, accidentally, or, in a sere old age, naturally? and how is it related to the rich green bosses that grow in deep velvet? And there again is another matter not clear to me. One calls them 'velvet' because they are all brought to an even surface at the top. Our own velvet is reduced to such trimness by cutting. But how is the moss trimmed? By what scissors? Carefullest Elizabethan gardener never shaped his yew hedge more daintily than the moss fairies smooth these soft rounded surfaces of green and gold. And just fancy the difference, if they were ragged! If the fibres had every one of them leave to grow at their own sweet will, and to be long or short as they liked, or, worse still, urged by fairy prizes into laboriously and agonizingly trying which could grow longest. Fancy the surface of a spot of competitive moss!

16. But how is it that they are subdued into that spherical obedience, like a crystal of wavellite?* Strange—that the vegetable creatures growing so fondly on rocks should form themselves in that mineral-like manner. It is true that the tops of all well-grown trees are rounded, on a large scale, as equally; but that is because they grow from a central stem, while these mossy mounds are made out of independent filaments, each growing to exactly his proper height in the sphere—short ones

* The reader should buy a small specimen of this mineral; it is a useful type of many structures.

outside, long in the middle. Stop, though; *is* that so? I am not even sure of that; perhaps they are built over a little dome of decayed moss below.* I must find out how every filament grows, separately—from root to cap, through the spirally set leaves. And meanwhile I don't know very clearly so much as what a root is—or what a leaf is. Before puzzling myself any further in examination either of moss or any other grander vegetable, I had better define these primal forms of all vegetation, as well as I can—or rather begin the definition of them, for future completion and correction. For, as my reader must already sufficiently perceive, this book is literally to be one of studies—not of statements. Some one said of me once, very shrewdly, When he wants to work out a subject, he writes a book on it. That is a very true saying in the main,—I work down or up to my mark, and let the reader see process and progress, not caring to conceal them. But this book will be nothing but process. I don't mean

* LUCCA, *Aug. 9th*, 1874.—I have left this passage as originally written, but I believe the dome is of accumulated earth. Bringing home, here, evening after evening, heaps of all kinds of mosses from the hills among which the Archbishop Ruggieri was hunting the wolf and her whelps in Ugolino's dream, I am more and more struck, every day, with their special function as earth-gatherers, and with the enormous importance to their own brightness, and to our service, of that dark and degraded state of the inferior leaves. And it fastens itself in my mind mainly as their distinctive character, that

to assert anything positively in it from the first page to the last. Whatever I say, is to be understood only as a conditional statement—liable to, and inviting, correction. And this the more because, as on the whole, I am at war with the botanists, I can't ask them to help me, and then call them names afterwards. I hope only for a contemptuous heaping of coals on my head by correction of my errors from them;—in some cases, my scientific friends will, I know, give me forgiving aid;—but, for many reasons, I am forced first to print the imperfect statement, as I can independently shape it; for if once I asked for, or received help, every thought would be frost-bitten into timid expression,

as the leaves of a tree become wood, so the leaves of a moss become earth, while yet a normal part of the plant. Here is a cake in my hand weighing half a pound, bright green on the surface, with minute crisp leaves; but an inch thick beneath in what looks at first like clay, but is indeed knitted fibre of exhausted moss. Also, I don't at all find the generalization I made from the botanical books likely to have occurred to me from the real things. No moss leaves that I can find here give me the idea of resemblance to pineapple leaves; nor do I see any, through my weak lens, clearly serrated; but I do find a general tendency to run into a silky filamentous structure, and in some, especially on a small one gathered from the fissures in the marble of the cathedral, white threads of considerable length at the extremities of the leaves, of which threads I remember no drawing or notice in the botanical books. Figure 1 represents, magnified, a cluster of these leaves, with the germinating stalk springing from their centre; but my scrawl was tired and careless, and for once Mr. Burgess has copied *too* accurately.

and every sentence broken by apology. I should have to write a dozen of letters before I could print a line, and the line, at last, would be only like a bit of any other botanical book—trustworthy it might be, perhaps; but certainly unreadable. Whereas now, it will rather put things more forcibly in the reader's mind to have them retouched and corrected as we go on; and our natural and honest mistakes will often be suggestive of things we could not have discovered but by wandering.

On these guarded conditions, then, I proceed to study, with my reader, the first general laws of vegetable form.

CHAPTER II.

THE ROOT.

1. PLANTS in their perfect form consist of four principal parts,—the Root, Stem, Leaf, and Flower. It is true that the stem and flower are parts, and remnants, or altered states, of the leaves; and that, speaking with close accuracy, we might say, a perfect plant consists of leaf and root. But the division into these four parts is best for practical purposes, and it will be desirable to note a few general facts about each, before endeavouring to describe any one kind of plant. Only, because the character of the stem depends on the nature of the leaf and flower, we must put it last in order of examination; and trace the development of the plant first in root and leaf; then in the flower and its fruit; and lastly in the stem.

2. First, then, the Root.

Every plant is divided, as I just said, in the main, into two parts, and these have opposite natures. One part seeks the light; the other hates it. One part feeds on the air; the other on the dust.

The part that loves the light is called the Leaf. It is an old Saxon word; I cannot get at its origin. The part that hates the light is called the Root.

In Greek, ῥίζα, Rhiza.*

In Latin, Radix, "the growing thing," which shortens, in French, into Race, and then they put on the diminutive 'ine,' and get their two words, Race, and Racine, of which we keep Race for animals, and use for vegetables a word of our own Saxon (and Dutch) dialect,—'root'; (connected with Rood—an image of wood; whence at last the Holy Rood, or Tree).

3. The Root has three great functions:

1st. To hold the plant in its place.

2nd. To nourish it with earth.

3rd. To receive vital power for it from the earth.

With this last office is in some degree,—and especially in certain plants,—connected, that of reproduction.

But in all plants the root has these three essential functions.

First, I said, to hold the Plant in its place. The Root is its Fetter.

* Learn this word, at any rate; and if you know any Greek, learn also this group of words: "ὡς ῥίζα ἐν γῇ διψώσῃ," which you may chance to meet with, and even to think about, some day.

You think it, perhaps, a matter of course that a plant is not to be a crawling thing? It is not a matter of course at all. A vegetable might be just what it is now, as compared with an animal;—might live on earth and water instead of on meat,—might be as senseless in life, as calm in death, and in all its parts and apparent structure unchanged; and yet be a crawling thing. It is quite as easy to conceive plants moving about like lizards, putting forward first one root and then another, as it is to think of them fastened to their place. It might have been well for them, one would have thought, to have the power of going down to the streams to drink, in time of drought;—of migrating in winter with grim march from north to south of Dunsinane Hill side. But that is not their appointed Fate. They are—at least, all the noblest of them—rooted to their spot. Their honour and use is in giving immoveable shelter,—in remaining landmarks, or lovemarks, when all else is changed:

"The cedars wave on Lebanon,
But Judah's statelier maids are gone."

4. Its root is thus a form of fate to the tree. It condemns, or indulges it, in its place. These semi-living creatures, come what may, shall abide, happy, or tormented. No doubt concerning "the position

in which Providence has placed *them*," is to trouble their minds, except so far as they can mend it by seeking light, or shrinking from wind, or grasping at support, within certain limits. In the thoughts of men they have thus become twofold images,—on the one side, of spirits restrained and half destroyed, whence the fables of transformation into trees; on the other, of spirits patient and continuing, having root in themselves and in good ground, capable of all persistent effort and vital stability, both in themselves, and for the human States they form.

5. In this function of holding fast, roots have a power of grasp quite different from that of branches. It is not a grasp, or clutch by contraction, as that of a bird's claw, or of the small branches we call 'tendrils' in climbing plants. It is a dead, clumsy, but inevitable grasp, by swelling, *after* contortion. For there is this main difference between a branch and root, that a branch cannot grow vividly but in certain directions and relations to its neighbour branches; but a root can grow wherever there is earth, and can turn in any direction to avoid an obstacle.*

* "Duhamel, botanist of the last century, tells us that, wishing to preserve a field of good land from the roots of an avenue of elms which were exhausting it, he cut a ditch between the field and avenue to

6. In thus contriving access for itself where it chooses, a root contorts itself into more serpent-like writhing than branches can; and when it has once coiled partly round a rock, or stone, it grasps it tight, necessarily, merely by swelling. Now a root has force enough sometimes to split rocks, but not to crush them; so it is compelled to grasp by *flattening* as it thickens; and, as it must have room somewhere, it alters its own shape as if it were made of dough, and holds the rock, not in a claw, but in a wooden cast or mould, adhering to its surface. And thus it not only finds its anchorage in the rock, but binds the rocks of its anchorage with a constrictor cable.

7. Hence—and this is a most important secondary function—roots bind together the ragged edges of rocks as a hem does the torn edge of a dress: they literally stitch the stones together; so that, while it is always dangerous to pass under a treeless edge of overhanging crag, as soon as it has become beautiful with trees, it is safe also. The rending power of roots on rocks has been greatly overrated. Capillary attraction in a willow wand

intercept the roots. But he saw with surprise those of the roots which had not been cut, go down behind the slope of the ditch to keep out of the light, go under the ditch, and into the field again." And the Swiss naturalist Bonnet said wittily, apropos of a wonder of this sort, "that sometimes it was difficult to distinguish a cat from a rose-bush."

will indeed split granite, and swelling roots sometimes heave considerable masses aside, but on the whole, roots, small and great, bind, and do not rend.* The surfaces of mountains are dissolved and disordered, by rain, and frost, and chemical decomposition, into mere heaps of loose stones on their desolate summits; but, where the forests grow, soil accumulates and disintegration ceases. And by cutting down forests on great mountain slopes, not only is the climate destroyed, but the danger of superficial landslip fearfully increased.

8. The second function of roots is to gather for the plant the nourishment it needs from the ground. This is partly water, mixed with some kinds of air (ammonia, etc.), but the plant can get both water and ammonia from the atmosphere; and, I believe, for the most part does so; though, when it cannot get water from the air, it will gladly drink by its roots. But the things it cannot receive from the air at all are certain earthy salts, essential to it (as iron is essential in our own blood), and of which, when it has quite exhausted the earth, no more such plants can grow in that ground. On this subject you will find enough in any modern

* As the first great office of the mosses is the gathering of earth, so that of the grasses is the binding of it. Theirs the Enchanter's toil, not in vain,—making ropes out of sea-sand.

treatise on agriculture; all that I want you to note here is that this feeding function of the root is of a very delicate and discriminating kind, needing much searching and mining among the dust, to find what it wants. If it only wanted water, it could get most of that by spreading in mere soft senseless limbs, like sponge, as far, and as far down, as it could; but to get the *salt* out of the earth it has to *sift* all the earth, and taste and touch every grain of it that it can, with fine fibres. And therefore a root is not at all a merely passive sponge or absorbing thing, but an infinitely subtle tongue, or tasting and eating thing. That is why it is always so fibrous and divided and entangled in the clinging earth.

9. "Always fibrous and divided"? But many roots are quite hard and solid!

No; the active part of the root is always, I believe, a fibre. But there is often a provident and passive part—a savings bank of root—in which nourishment is laid up for the plant, and which, though it may be underground, is no more to be considered its real root than the kernel of a seed is. When you sow a pea, if you take it up in a day or two, you will find the fibre below, which is root; the shoot above, which is plant; and the pea as a now partly exhausted storehouse, looking very woful, and like the granaries of Paris after the fire. So, the round solid root of a cyclamen, or the conical one which you know so well as a carrot, are not properly roots, but permanent storehouses,—only the fibres that grow from them are roots. Then there are other apparent roots which are not even storehouses, but refuges; houses where the little plant lives in its infancy, through winter and rough weather. So that it will be best for you at once to limit your idea of a root to this,—that it is a group of growing fibres which taste and suck what is good for the plant out of the ground, and by their united strength hold it in its place; only remember the thick limbs of roots do not feed, but only the fine fibres at the ends of them which are something between tongues and sponges, and while they absorb moisture readily, are yet as particular about getting what they think nice to eat as any dainty little boy or girl; looking for it everywhere, and turning angry and sulky if they don't get it.

10. But the root has, it seems to me, one more function, the most important of all. I say, it seems to me, for observe, what I have hitherto told you is all (I believe) ascertained and admitted; this that I am going to tell you has not yet, as far as I know, been asserted by men of science, though I believe it

to be demonstrable. But you are to examine into it, and think of it for yourself.

There are some plants which appear to derive all their food from the air—which need nothing but a slight grasp of the ground to fix them in their place. Yet if we were to tie them into that place, in a framework, and cut them from their roots, they would die. Not only in these, but in all other plants, the vital power by which they shape and feed themselves, whatever that power may be, depends, I think, on that slight touch of the earth, and strange inheritance of its power. It is as essential to the plant's life as the connection of the head of an animal with its body by the spine is to the animal. Divide the feeble nervous thread, and all life ceases. Nay, in the tree the root is even of greater importance. You will not kill the tree, as you would an animal, by dividing its body or trunk. The part not severed from the root will shoot again. But in the root, and its touch of the ground, is the life of it. My own definition of a plant would be "a living creature whose source of vital energy is in the earth" (or in the water, as a form of the earth; that is, in inorganic substance). There is, however, one tribe of plants which seems nearly excepted from this law. It is a very strange one, having long been noted for the resemblance of its flowers to different insects; and it has recently been proved by Mr. Darwin to be dependent on insects for its existence. Doubly strange therefore, it seems, that in some cases this race of plants all but reaches the independent life of insects. It rather *settles* upon boughs than roots itself in them; half of its roots may wave in the air.

11. What vital power is, men of science are not a step nearer knowing than they were four thousand years ago. They are, if anything, farther from knowing now than then, in that they imagine themselves nearer. But they know more about its limitations and manifestations than they did. They have even arrived at something like a proof that there is a fixed quantity of it flowing out of things and into them. But, for the present, rest content with the general and sure knowledge that, fixed or flowing, measurable or immeasurable—one with electricity or heat or light, or quite distinct from any of them—life is a delightful, and its negative death, a dreadful thing, to human creatures; and that you can give or gather a certain quantity of life into plants, animals, and yourself by wisdom and courage, and by their reverses can bring upon them any quantity of death you please, which is a much more serious point for you to consider than what life and death are.

12. Now, having got a quite clear idea of a root properly so called, we may observe what those storehouses, refuges, and ruins are, which we find connected with roots. The greater number of plants feed and grow at the same time; but there are some of them which like to feed first and grow afterwards. For the first year, or, at all events, the first period of their life, they gather material for their future life out of the ground and out of the air, and lay it up in a storehouse, as bees make combs. Of these stores—for the most part rounded masses tapering downwards into the ground—some are as good for human beings as honeycombs are; only not so sweet. We steal them from the plants, as we do from the bees, and these conical upside-down hives or treasuries of Atreus, under the names of carrots, turnips, and radishes, have had important influence on human fortunes. If we do not steal the store, next year the plant lives upon it, raises its stem, flowers and seeds out of that abundance, and having fulfilled its destiny, and provided for its successor, passes away, root and branch together.

13. There is a pretty example of patience for us in this; and it would be well for young people generally to set themselves to grow in a carrotty or turnippy manner, and lay up secret store, not caring

to exhibit it until the time comes for fruitful display. But they must not, in after-life, imitate the spendthrift vegetable, and blossom only in the strength of what they learned long ago; else they soon come to contemptible end. Wise people live like laurels and cedars, and go on mining in the earth, while they adorn and embalm the air.

14. Secondly, Refuges. As flowers growing on trees have to live for some time, when they are young, in their buds, so some flowers growing on the ground have to live for a while, when they are young, *in* what we call their roots. These are mostly among the Drosidæ * and other humble tribes, loving the ground; and, in their babyhood, liking to live quite down in it. A baby crocus has literally its own little dome—domus, or duomo—within which in early spring it lives a delicate convent life of its own, quite free from all worldly care and dangers, exceedingly ignorant of things in general, but itself brightly golden and perfectly formed before it is brought out. These subterranean palaces and vaulted cloisters, which we call bulbs, are no more roots than the blade of grass is a root, in which the ear of corn forms before it shoots up.

* Drosidæ, in our school nomenclature, is the general name, including the four great tribes, iris, asphodel, amaryllis, and lily. See reason for this name given in the 'Queen of the Air,' Section II.

15. Thirdly, Ruins. The flowers which have these subterranean homes form one of many families whose roots, as well as seeds, have the power of reproduction. The succession of some plants is trusted much to their seeds: a thistle sows itself by its down, an oak by its acorns; the companies of flying emigrants settle where they may; and the shadowy tree is content to cast down its showers of nuts for swines' food with the chance that here and there one may become a ship's bulwark. But others among plants are less careless, or less proud. Many are anxious for their children to grow in the place where they grew themselves, and secure this not merely by letting their fruit fall at their feet, on the chance of its growing up beside them, but by closer bond, bud springing fórth from root, and the young plant being animated by the gradually surrendered life of its parent. Sometimes the young root is formed above the old one, as in the crocus, or beside it, as in the amaryllis, or beside it in a spiral succession, as in the orchis; in these cases the old root always perishes wholly when the young one is formed; but in a far greater number of tribes, one root connects itself with another by a short piece of intermediate stem; and this stem does not at once perish when the new root is formed, but grows on at one end indefi-

nitely, perishing slowly at the other, the scars or ruins of the past plants being long traceable on its sides. When it grows entirely underground it is called a root-stock. But there is no essential distinction between a root-stock and a creeping stem, only the root-stock may be thought of as a stem which shares the melancholy humour of a root in loving darkness, while yet it has enough consciousness of better things to grow towards, or near, the light. In one family it is even fragrant where the flower is not, and a simple houseleek is called 'rhodiola rosea,' because its root-stock has the scent of a rose.

16. There is one very unusual condition of the root-stock which has become of much importance in economy, though it is of little in botany; the forming, namely, of knots at the ends of the branches of the underground stem, where the new roots are to be thrown out. Of these knots, or 'tubers,' (swollen things,) one kind, belonging to the tobacco tribe, has been singularly harmful, together with its pungent relative, to a neighbouring country of ours, which perhaps may reach a higher destiny than any of its friends can conceive for it, if it can ever succeed in living without either the potato, or the pipe.

17. Being prepared now to find among plants

many things which are like roots, yet are not, you may simplify and make fast your true idea of a root as a fibre or group of fibres, which fixes, animates, and partly feeds the leaf. Then practically, as you examine plants in detail, ask first respecting them: What kind of root have they? Is it large or small in proportion to their bulk, and why is it so? What soil does it like, and what properties does it acquire from it? The endeavour to answer these questions will soon lead you to a rational inquiry into the plant's history. You will first ascertain what rock or earth it delights in, and what climate and circumstances; then you will see how its root is fitted to sustain it mechanically under given pressures and violences, and to find for it the necessary sustenance under given difficulties of famine or drought. Lastly you will consider what chemical actions appear to be going on in the root, or its store; what processes there are, and elements, which give pungency to the radish, flavour to the onion, or sweetness to the liquorice; and of what service each root may be made capable under cultivation, and by proper subsequent treatment, either to animals or men.

18. I shall not attempt to do any of this for you; I assume, in giving this advice, that you wish to pursue the science of botany as your chief study; I

have only broken moments for it, snatched from my chief occupations, and I have done nothing myself of all this I tell you to do. But so far as you can work in this manner, even if you only ascertain the history of one plant, so that you know that accurately, you will have helped to lay the foundation of a true science of botany, from which the mass of useless nomenclature,* now mistaken for science, will fall away, as the husk of a poppy falls from the bursting flower.

* The only use of a great part of our existing nomenclature is to enable one botanist to describe to another, a plant which the other has not seen. When the science becomes approximately perfect, all known plants will be properly figured, so that nobody need describe them; and unknown plants be so rare that nobody will care to learn a new and difficult language, in order to be able to give an account of what in all probability he will never see.

Drawn by J Ruskin.

Engraved by G Allen

II.

Central Type of Leaves.

COMMON BAY-LAUREL.

CHAPTER III.

THE LEAF.

1. IN the first of the poems of which the English Government has appointed a portion to be sung every day for the instruction and pleasure of the people, there occurs this curious statement respecting any person who will behave himself rightly: "He shall be like a tree planted by the river side, that bears its fruit in its season. His leaf also shall not wither; and you will see that whatever he does will prosper."

I call it a curious statement, because the conduct to which this prosperity is promised is not that which the English, as a nation, at present think conducive to prosperity: but whether the statement be true or not, it will be easy for you to recollect the two eastern figures under which the happiness of the man is represented,—that he is like a tree bearing fruit "in its season"; (not so hastily as that the frost pinch it, nor so late that no sun ripens it;) and that "his leaf shall not fade." I should like you to recollect this phrase in the

Vulgate—"folium ejus non defluet"—shall not fall *away*,—that is to say, shall not fall so as to leave any visible bareness in winter time, but only that others may come up in its place, and the tree be always green.

2. Now, you know, the fruit of the tree is either for the continuance of its race, or for the good, or harm, of other creatures. In no case is it a good to the tree itself. It is not indeed, properly, a part of the tree at all, any more than the egg is part of the bird, or the young of any creature part of the creature itself. But in the leaf is the strength of the tree itself. Nay, rightly speaking, the leaves *are* the tree itself. Its trunk sustains; its fruit burdens and exhausts; but in the leaf it breathes and lives. And thus also, in the eastern symbolism, the fruit is the labour of men for others; but the leaf is their own life. "He shall bring forth fruit, in his time; and his own joy and strength shall be continual."

3. Notice next the word 'folium.' In Greek, φυλλον, 'phyllon.'

"The thing that is born," or "put forth." "When the branch is tender, and putteth forth her leaves, ye know that summer is nigh." The botanists say, "The leaf is an expansion of the bark of the stem." More accurately, the bark is a contraction of the tissue of the leaf. For every leaf is born out of the earth, and breathes out of the air; and there are many leaves that have no stems, but only roots. It is 'the springing thing'; this thin film of life; rising, with its *edge* out of the ground—infinitely feeble, infinitely fair. With Folium, in Latin, is rightly associated the word Flos; for the flower is only a group of singularly happy leaves. From these two roots come foglio, feuille, feuillage, and fleur;—blume, blossom, and bloom; our foliage, and the borrowed foil, and the connected technical groups of words in architecture and the sciences.

4. This *thin* film, I said. That is the essential character of a leaf; to be thin,—widely spread out in proportion to its mass. It is the opening of the substance of the earth to the air, which is the giver of life. The Greeks called it, therefore, not only the born or blooming thing, but the spread or expanded thing—"πέταλον." Pindar calls the beginnings of quarrel, "petals of quarrel." Recollect, therefore, this form, Petalos; and connect it with Petasos, the expanded cap of Mercury. For one great use of both is to give shade. The root of all these words is said to be ΠΕΤ (Pet), which may easily be remembered in Greek, as it sometimes occurs in no unpleasant sense in English.

5. But the word 'petalos' is connected in Greek with another word, meaning, to fly,—so that you may think of a bird as spreading its petals to the wind; and with another, signifying Fate in its pursuing flight, the overtaking thing, or over-flying Fate. Finally, there is another Greek word meaning 'wide,' πλατυς (platys); whence at last our 'plate'—a thing made broad or extended—but especially made broad or 'flat' out of the solid, as in a lump of clay extended on the wheel, or a lump of metal extended by the hammer. So the first we call Platter; the second Plate, when of the precious metals. Then putting *b* for *p*, and *d* for *t*, we get the blade of an oar, and blade of grass.

6. Now gather a branch of laurel, and look at it carefully. You may read the history of the being of half the earth in one of those green oval leaves—the things that the sun and the rivers have made out of dry ground. Daphne—daughter of Enipeus, and beloved by the Sun,—that fable gives you at once the two great facts about vegetation. Where warmth is, and moisture—there, also, the leaf. Where no warmth—there is no leaf; where there is no dew—no leaf.

7. Look, then, to the branch you hold in your hand. That you *can* so hold it, or make a crown

of it, if you choose, is the first thing I want you to note of it;—the proportion of size, namely, between the leaf and *you*. Great part of your life and character, as a human creature, has depended on that. Suppose all leaves had been spacious, like some palm leaves; solid, like cactus stem; or that trees had grown, as they might of course just as easily have grown, like mushrooms, all one great cluster of leaf round one stalk. I do not say that they are divided into small leaves only for your delight, or your service, as if you were the monarch of everything—even in this atom of a globe. You are made of your proper size; and the leaves of theirs: for reasons, and by laws, of which neither the leaves nor you know anything. Only note the harmony between both, and the joy we may have in this division and mystery of the frivolous and tremulous petals, which break the light and the breeze,—compared to what, with the frivolous and tremulous mind which is in us, we could have had out of domes, or penthouses, or walls of leaf.

8. Secondly; think awhile of its dark clear green, and the good of it to you. Scientifically, you know green in leaves is owing to 'chlorophyll,' or, in English, to 'green-leaf.' It may be very fine to know that; but my advice to you, on the whole, is to rest content with the general fact that leaves

are green when they do not grow in or near smoky towns; and not by any means to rest content with the fact that very soon there will not be a green leaf in England, but only greenish-black ones. And thereon resolve that you will yourself endeavour to promote the growing of the green wood, rather than of the black.

9. Looking at the back of your laurel-leaves, you see how the central rib or spine of each, and the lateral branchings, strengthen and carry it. I find much confused use, in botanical works, of the words Vein and Rib. For, indeed, there are veins *in* the ribs of leaves, as marrow in bones; and the projecting bars often gradually depress themselves into a transparent net of rivers. But the *mechanical* force of the framework in carrying the leaf-tissue is the point first to be noticed; it is that which admits, regulates, or restrains the visible motions of the leaf; while the system of circulation can only be studied through the microscope. But the ribbed leaf bears itself to the wind, as the webbed foot of a bird does to the water, and needs the same kind, though not the same strength, of support; and its ribs always are partly therefore constituted of strong woody substance, which is knit out of the tissue; and you can extricate this skeleton framework, and keep it, after the leaf-tissue

is dissolved. So I shall henceforward speak simply of the leaf and its ribs,—only specifying the additional veined structure on necessary occasions.

10. I have just said that the ribs—and might have said, farther, the stalk that sustains them—are knit out of the *tissue* of the leaf. But what is the leaf-tissue itself knit out of? One would think that was nearly the first thing to be discovered, or at least to be thought of, concerning plants,—namely, how and of what they are made. We say they 'grow.' But you know that they can't grow out of nothing;—this solid wood and rich tracery must be made out of some previously existing substance. What is the substance?—and how is it woven into leaves,—twisted into wood?

11. Consider how fast this is done, in spring. You walk in February over a slippery field, where, through hoar-frost and mud, you perhaps hardly see the small green blades of trampled turf. In twelve weeks you wade through the same field up to your knees in fresh grass; and in a week or two more, you mow two or three solid haystacks off it. In winter you walk by your currant-bush, or your vine. They are shrivelled sticks—like bits of black tea in the canister. You pass again in May, and the currant-bush looks like a young sycamore tree; and the vine is a bower: and meanwhile the forests,

all over this side of the round world, have grown their foot or two in height, with new leaves—so much deeper, so much denser than they were. Where has it all come from? Cut off the fresh shoots from a single branch of any tree in May. Weigh them; and then consider that so much weight has been added to every such living branch, everywhere, this side the equator, within the last two months. What is all that made of?

12. Well, this much the botanists really know, and tell us,—It is made chiefly of the breath of animals: that is to say, of the substance which, during the past year, animals have breathed into the air; and which, if they went on breathing, and their breath were not made into trees, would poison them, or rather suffocate them, as people are suffocated in uncleansed pits, and dogs in the Grotta del Cane. So that you may look upon the grass and forests of the earth as a kind of green hoar-frost, frozen upon it from our breath, as, on the window-panes, the white arborescence of ice.

13. But how is it made into wood?

The substances that have been breathed into the air are charcoal, with oxygen and hydrogen,—or, more plainly, charcoal and water. Some necessary earth,—in smaller quantity, but absolutely essential,—the trees get from the ground; but, I believe all the charcoal they want, and most of the water, from the air. Now the question is, where and how do they take it in, and digest it into wood?

14. You know, in spring, and partly through all the year, except in frost, a liquid called 'sap' circulates in trees, of which the nature, one should have thought, might have been ascertained by mankind in the six thousand years they have been cutting wood. Under the impression always that it *had been* ascertained, and that I could at any time know all about it, I have put off till to-day, 19th October, 1869, when I am past fifty, the knowing anything about it at all. But I will really endeavour now to ascertain something, and take to my botanical books, accordingly, in due order.

(1) Dresser's "Rudiments of Botany." 'Sap' not in the index; only Samara, and Sarcocarp,—about neither of which I feel the smallest curiosity. (2) Figuier's "Histoire des Plantes."* 'Sêve,' not in index; only Serpolet, and Sherardia arvensis, which also have no help in them for me. (3) Balfour's "Manual of Botany." 'Sap,'—yes, at last. "Article 257. Course of fluids in exogenous stems." I don't care about the course just now: I want to know where the fluids come from. "If a plant be

* An excellent book, nevertheless.

plunged into a weak solution of acetate of lead,"—I don't in the least want to know what happens. "From the minuteness of the tissue, it is not easy to determine the vessels through which the sap moves." Who said it was? If it had been easy, I should have done it myself. "Changes take place in the composition of the sap in its upward course." I dare say; but I don't know yet what its composition is before it begins going up. "The Elaborated Sap by Mr. Schultz has been called 'latex.'" I wish Mr. Schultz were in a hogshead of it, with the top on. "On account of these movements in the latex, the laticiferous vessels have been denominated cinenchymatous." I do not venture to print the expressions which I here mentally make use of.

15. Stay,—here, at last, in Article 264, is something to the purpose: "It appears then that, in the case of Exogenous plants, the fluid matter in the soil, containing different substances in solution, is sucked up by the extremities of the roots." Yes, but how of the pine trees on yonder rock?—Is there any sap in the rock, or water either? The moisture must be seized during actual rain on the root, or stored up from the snow; stored up, any way, in a tranquil, not actively sappy, state, till the time comes for its change, of which there is no account here.

16. I have only one chance left now. Lindley's "Introduction to Botany." 'Sap,'—yes,—'General motion of.' II. 325. "The course which is taken by the sap, after entering a plant, is the first subject for consideration." My dear Doctor, I have learned nearly whatever I know of plant structure from you, and am grateful; and that it is little, is not your fault, but mine. But this—let me say it with all sincere respect—is not what you should have told me here. You know, far better than I, that 'sap' never does enter a plant at all; but only salt, or earth and water, and that the roots alone could not make it; and that, therefore, the course of it must be, in great part, the result or process of the actual making. But I will read now, patiently; for I know you will tell me much that is worth hearing, though not perhaps what I want.

Yes; now that I have read Lindley's statement carefully, I find it is full of precious things; and this is what, with thinking over it, I can gather for you.

17. First, towards the end of January,—as the light enlarges, and the trees revive from their rest,—there is a general liquefaction of the blood of St. Januarius in their stems; and I suppose there is really a great deal of moisture rapidly absorbed from the earth in most cases; and that this absorption is a great help

to the sun in drying the winter's damp out of it for us: then, with that strange vital power,—which scientific people are usually as afraid of naming as common people are afraid of naming Death,—the tree gives the gathered earth and water a changed existence; and to this new-born liquid an upward motion from the earth, as our blood has from the heart; for the life of the tree is out of the earth; and this upward motion has a mechanical power in pushing on the growth. "*Forced onward* by the current of sap, the plumule ascends," (Lindley, p. 132,)—this blood of the tree having to supply, exactly as our own blood has, not only the forming powers of substance, but a continual evaporation, "approximately seventeen times more than that of the human body," while the force of motion in the sap "is sometimes five times greater than that which impels the blood in the crural artery of the horse."

18. Hence generally, I think we may conclude thus much,—that at every pore of its surface, under ground and above, the plant in the spring absorbs moisture, which instantly disperses itself through its whole system "by means of some permeable quality of the membranes of the cellular tissue invisible to our eyes even by the most powerful glasses" (p. 326); that in this way subjected to the vital

power of the tree, it becomes sap, properly so called, which passes downwards through this cellular tissue, slowly and secretly; and then upwards, through the great vessels of the tree, violently, stretching out the supple twigs of it as you see a flaccid waterpipe swell and move when the cock is turned to fill it. And the tree becomes literally a fountain, of which the springing streamlets are clothed with new-woven garments of green tissue, and of which the silver spray stays in the sky,—a spray, now, of leaves.

19. That is the gist of the matter; and a very wonderful gist it is, to my mind. The secret and subtle descent—the violent and exulting resilience of the tree's blood,—what guides it?—what compels? The creature has no heart to beat like ours; one cannot take refuge from the mystery in a 'muscular contraction.' Fountain without supply—playing by its own force, for ever rising and falling all through the days of Spring, spending itself at last in gathered clouds of leaves, and iris of blossom.

Very wonderful; and it seems, for the present, that we know nothing whatever about its causes;—nay, the strangeness of the reversed arterial and vein motion, without a heart, does not seem to strike anybody. Perhaps, however, it may interest you, as I observe it does the botanists, to know that the cellular tissue through which the motion effected is called Parenchym, and the woody tissue, Bothrenchym; and that Parenchym is divided, by a system of nomenclature which "has some advantages over that more commonly in use,"* into merenchyma, conenchyma, ovenchyma, atractenchyma, cylindrenchyma, colpenchyma, cladenchyma, and prismenchyma.

20. Take your laurel branch into your hand again. There are, as you must well know, innumerable shapes and orders of leaves;—there are some like paws, and some like claws; some like fingers, and some like feet; there are endlessly cleft ones, and endlessly clustered ones, and inscrutable divisions within divisions of the fretted verdure; and wrinkles, and ripples, and stitchings, and hemmings, and pinchings, and gatherings, and crumplings, and clippings, and what not. But there is nothing so constantly noble as the pure leaf of the laurel, bay, orange, and olive; numerable, sequent, perfect in setting, divinely simple and serene. I shall call these noble leaves 'Apolline' leaves. They characterize many orders of plants, great and small,—from the magnolia to the myrtle, and exquisite 'myrtille' of the hills (bilberry); but wherever you find them, strong, lustrous, dark green, simply

* Lindley, 'Introduction to Botany,' vol. i., p. 21. The terms "wholly obsolete," says an authoritative botanical friend. Thank Heaven

formed, richly scented or stored,—you have nearly always kindly and lovely vegetation, in healthy ground and air.

21. The gradual diminution in rank beneath the Apolline leaf, takes place in others by the loss of one or more of the qualities above named. The Apolline leaf, I said, is strong, lustrous, full in its green, rich in substance, simple in form. The inferior leaves are those which have lost strength, and become thin, like paper; which have lost lustre, and become dead by roughness of surface, like the nettle,—(an Apolline leaf may become dead by *bloom*, like the olive, yet not lose beauty); which have lost colour, and become feeble in green, as in the poplar, or *crudely* bright, like rice; which have lost substance and softness, and have nothing to give in scent or nourishment; or become flinty or spiny; finally, which have lost simplicity, and become cloven or jagged. Many of these losses are partly atoned for by gain of some peculiar loveliness. Grass and moss, and parsley and fern, have each their own delightfulness; yet they are all of inferior power and honour, compared to the Apolline leaves.

22. You see, however, that though your laurel leaf has a central stem, and traces of ribs branching from it, in a vertebrated manner, they are so

faint that we cannot take it for a type of vertebrate structure. But the two figures of elm and alisma leaf, given in "Modern Painters" (vol. iii.), and now here repeated, Fig. 3, will clearly enough show the opposition between this vertebrate form, branching again usually at the edges, *a*, and the softly opening lines diffused at the stem, and gathered at the point of the leaf, *b*, which, as you almost

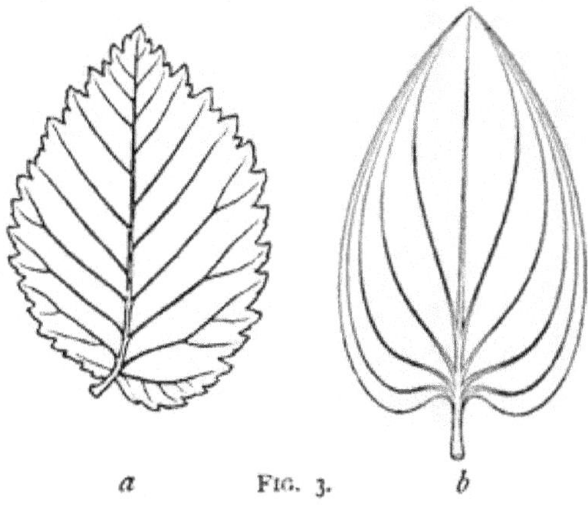

a Fig. 3. *b*

without doubt know already, are characteristic of a vast group of plants, including especially all the lilies, grasses, and palms, which for the most part are the signs of local or temporary moisture in hot countries;—local, as of fountains and streams; temporary, as of rain or inundation.

But temporary, still more definitely in the day, than in the year. When you go out, delighted, into the dew of the morning, have you ever considered why it is so rich upon the grass;—why it is *not* upon the trees? It *is* partly on the trees, but yet your memory of it will be always chiefly of its gleam upon the lawn. On many trees you will find there is none at all. I cannot follow out here the many inquiries connected with this subject, but, broadly, remember the branched trees are fed chiefly by rain,—the unbranched ones by dew, visible or invisible; that is to say, at all events by moisture which they can gather for themselves out of the air; or else by streams and springs. Hence the division of the verse of the song of Moses: "My doctrine shall drop as the rain; my speech shall distil as the dew: as the *small* rain upon the tender *herb*, and as the showers upon the grass."

23. Next, examining the direction of the veins in the leaf of the alisma, *b*, Fig. 3, you see they all open widely, as soon as they can, towards the thick part of the leaf; and then taper, apparently with reluctance, pushing each other outwards, to the point. If the leaf were a lake of the same shape, and its stem the entering river, the lines of the currents passing through it would, I believe, be nearly the same as that of the veins in the aquatic leaf. I have not examined the fluid law accurately, and I do not suppose there is more real correspondence

than may be caused by the leaf's expanding in every permitted direction, as the water would, with all the speed it can; but the resemblance is so close as to enable you to fasten the relation of the unbranched leaves to streams more distinctly in your mind,—just as the toss of the palm leaves from their stem may, I think, in their likeness to the springing of a fountain, remind you of their relation to the desert, and their necessity, therein, to life of man and beast.

24. And thus, associating these grass and lily leaves always with fountains, or with dew, I think we may get a pretty general name for them also. You know that Cora, our Madonna of the flowers, was lost in Sicilian Fields: you know, also, that the fairest of Greek fountains, lost in Greece, was thought to rise in a Sicilian islet; and that the real springing of the noble fountain in that rock was one of the causes which determined the position of the greatest Greek city of Sicily. So I think, as we call the fairest branched leaves 'Apolline,' we will call the fairest flowing ones 'Arethusan.' But remember that the Apolline leaf represents only the central type of land leaves, and is, within certain limits, of a fixed form; while the beautiful Arethusan leaves, alike in flowing of their lines, change their forms indefinitely,—some shaped like round pools,

and some like winding currents, and many like arrows, and many like hearts, and otherwise varied and variable, as leaves ought to be,—that rise out of the waters, and float amidst the pausing of their foam.

25. Brantwood, *Easter Day*, 1875.—I don't like to spoil my pretty sentence, above; but on reading it over, I suspect I wrote it confusing the water-lily leaf, and other floating ones of the same kind, with the Arethusan forms. But the water-lily and water-ranunculus leaves, and such others, are to the orders of earth-loving leaves what ducks and swans are to birds; (the swan is the water-lily of birds;) they are *swimming* leaves; not properly watery-creatures, or able to live under water like fish, (unless when dormant), but just like birds that pass their lives on the surface of the waves—though they must breathe in the air.

And these natant leaves, as they lie on the water surface, do not want strong ribs to carry them,* but have very delicate ones beautifully branching into the orbed space, to keep the tissue nice and flat; while, on the other hand, leaves that really have to grow under water, sacrifice their tissue, and keep only their ribs, like coral animals; ('Ranunculus heterophyllus,' 'other-leaved Frog-flower,' and its like,) just as, if

* "You should see the girders on under-side of the Victoria Water-lily, the most wonderful bit of engineering, of the kind, I know of."—('Botanical friend.')

you keep your own hands too long in water, they shrivel at the finger-ends.

26. So that you must not attach any great botanical importance to the characters of contrasted aspects in leaves, which I wish you to express by the words 'Apolline' and 'Arethusan'; but their mythic importance is very great, and your careful observance of it will help you completely to understand the beautiful Greek fable of Apollo and Daphne. There are indeed several Daphnes, and the first root of the name is far away in another field of thought altogether, connected with the Gods of Light. But etymology, the best of servants, is an unreasonable master; and Professor Max Müller trusts his deep-reaching knowledge of the first ideas connected with the names of Athena and Daphne, too implicitly, when he supposes this idea to be retained in central Greek theology. Athena' originally meant only the dawn, among nations who knew nothing of a Sacred Spirit. But the Athena who catches Achilles by the hair, and urges the spear of Diomed, has not, in the mind of Homer, the slightest remaining connection with the mere beauty of daybreak. Daphne chased by Apollo, may perhaps—though I doubt even this much of consistence in the earlier myth—have meant the Dawn pursued by the Sun. But there is

no trace whatever of this first idea left in the fable of Arcadia and Thessaly.

27. The central Greek Daphne is the daughter of one of the great *river* gods of Arcadia; her mother is the Earth. Now Arcadia is the Oberland of Greece; and the crests of Cyllene, Erymanthus, and Mænalus* surround it, like the Swiss forest cantons, with walls of rock, and shadows of pine. And it divides itself, like the Oberland, into three regions: first, the region of rock and snow, sacred to Mercury and Apollo, in which Mercury's birth on Cyllene, his construction of the lyre, and his stealing the oxen of Apollo, are all expressions of the enchantments of cloud and sound, mingling with the sunshine, on the cliffs of Cyllene.

> "While the mists
> Flying, and rainy vapours, call out shapes
> And phantoms from the crags and solid earth
> As fast as a musician scatters sounds
> Out of his instrument."

Then came the pine region, sacred especially to Pan and Mænalus, the son of Lycaon and brother of Callisto; and you had better remember this relationship carefully, for the sake of the meaning of the constellations of Ursa Major and the Mons Mænalius,

* Roughly, Cyllene 7,700 feet high; Erymanthus 7,000; Mænalus 6,000.

and of their wolf and bear traditions; (compare also the strong impression on the Greek mind of the wild leafiness, nourished by snow, of the Bœotian Cithæron, —"Oh, thou lake-hollow, full of divine leaves, and of wild creatures, nurse of the snow, darling of Diana," (Phœnissæ, 801). How wild the climate of this pine region is, you may judge from the pieces in the note below* out of Colonel Leake's diary in

* *March 3rd.*—We now ascend the roots of the mountain called Kastaniá, and begin to pass between it and the mountain of Alonístena, which is on our right. The latter is much higher than Kastaniá, and, like the other peaked summits of the Mænalian range, is covered with firs, and deeply at present with snow. The snow lies also in our pass. At a fountain in the road, the small village of Bazeniko is half a mile on the right, standing at the foot of the Mænalian range, and now covered with snow.

Saetá is the most lofty of the range of mountains, which are in face of Levídhi, to the northward and eastward; they are all a part of the chain which extends from Mount Khelmós, and connects that great summit with Artemisium, Parthenium, and Parnon. Mount Saetá is covered with firs. The mountain between the plain of Levídhi and Alonístena, or, to speak by the ancient nomenclature, that part of the Mænalian range which separates the Orchomenia from the valleys of Helisson and Methydrium, is clothed also with large forests of the same trees; the road across this ridge from Levídhi to Alonístena is now impracticable on account of the snow.

I am detained all day at Levídhi by a heavy fall of snow, which before the evening has covered the ground to half a foot in depth, although the village is not much elevated above the plain, nor in a more lofty situation than Tripolitzá.

March 4th.—Yesterday afternoon and during the night the snow fell in such quantities as to cover all the plains and adjacent mountains; and the country exhibited this morning as fine a snow-scene as Norway could supply. As the day advanced and the sun appeared, the snow melted rapidly, but the sky was soon overcast again, and the snow began to fall.

crossing the Mænalian range in spring. And then, lastly, you have the laurel and vine region, full of sweetness and Elysian beauty.

28. Now as Mercury is the ruling power of the hill enchantment, so Daphne of the leafy peace. She is, in her first life, the daughter of the mountain river, the mist of it filling the valley; the Sun, pursuing, and effacing it, from dell to dell, is, literally, Apollo pursuing Daphne, and *adverse* to her; (not, as in the earlier tradition, the Sun pursuing only his own light). Daphne, thus hunted, cries to her mother, the Earth, which opens, and receives her, causing the laurel to spring up in her stead. That is to say, wherever the rocks protect the mist from the sunbeam, and suffer it to water the earth, there the laurel and other richest vegetation fill the hollows, giving a better glory to the sun itself. For sunshine, on the torrent spray, on the grass of its valley, and entangled among the laurel stems, or glancing from their leaves, became a thousandfold lovelier and more sacred than the same sunbeams, burning on the leafless mountain-side.

And farther, the leaf, in its connection with the river, is typically expressive, not, as the flower was, of human fading and passing away, but of the perpetual flow and renewal of human mind and thought, rising "like the rivers that run among the hills";

therefore it was that the youth of Greece sacrificed their hair—the sign of their continually renewed strength,—to the rivers, and to Apollo. Therefore, to commemorate Apollo's own chief victory over death —over Python, the corrupter,—a laurel branch was gathered every ninth year in the vale of Tempe; and the laurel leaf became the reward or crown of all beneficent and enduring work of man—work of inspiration, born of the strength of the earth, and of the dew of heaven, and which can never pass away.

29. You may doubt at first, even because of its grace, this meaning in the fable of Apollo and Daphne; you will not doubt it, however, when you trace it back to its first eastern origin. When we speak carelessly of the traditions respecting the Garden of Eden, (or in Hebrew, remember, Garden of Delight,) we are apt to confuse Milton's descriptions with those in the book of Genesis. Milton fills his Paradise with flowers; but no flowers are spoken of in Genesis. We may indeed conclude that in speaking of every herb of the field, flowers are included. But they are not named. The things that are *named* in the Garden of Delight are trees only.

The words are, "every tree that was pleasant to the sight and good for food;" and as if to mark the idea more strongly for us in the Septuagint, even the ordinary Greek word for tree is not used, but the word ξυλον,—literally, every 'wood,' every piece of *timber* that was pleasant or good. They are indeed the "vivi travi,"—living rafters,—of Dante's Apennine.

Do you remember how those trees were said to be watered? Not by the four rivers only. The rivers could not supply the place of rain. No rivers do; for in truth they are the refuse of rain. No storm-clouds were there, nor hidings of the blue by darkening veil; but there went up a *mist* from the earth, and watered the face of the ground,—or, as in Septuagint and Vulgate, "There went forth a fountain from the earth, and gave the earth to drink."

30. And now, lastly, we continually think of that Garden of Delight, as if it existed, or could exist, no longer; wholly forgetting that it is spoken of in Scripture as perpetually existent; and some of its fairest trees as existent also, or only recently destroyed. When Ezekiel is describing to Pharaoh the greatness of the Assyrians, do you remember what image he gives of them? "Behold, the Assyrian was a cedar in Lebanon, with fair branches; and his top was among the thick boughs; the waters nourished him, and the deep brought him up, with her rivers

running round about his plants. Under his branches did all the beasts of the field bring forth their young; and under his shadow dwelt all great nations."

31. Now hear what follows. "The cedars *in the Garden of God* could not hide *him*. The fir trees were not like his boughs, and the chestnut trees were not like his branches: nor any tree in the Garden of God was like unto him in beauty."

So that you see, whenever a nation rises into consistent, vital, and, through many generations, enduring power, *there* is still the Garden of God; still it is the water of life which feeds the roots of it; and still the succession of its people is imaged by the perennial leafage of trees of Paradise. Could this be said of Assyria, and shall it not be said of England? How much more, of lives such as ours should be,—just, laborious, united in aim, beneficent in fulfilment,—may the image be used of the leaves of the trees of Eden! Other symbols have been given often to show the evanescence and slightness of our lives—the foam upon the water, the grass on the housetop, the vapour that vanishes away; yet none of these are images of true human life. That life, when it is real, is *not* evanescent; is *not* slight; does *not* vanish away. Every noble life leaves the fibre of it interwoven for ever in the work of the world; by so much, evermore, the strength of the human race has gained; more stubborn in the root, higher towards heaven in the branch; and, "as a teil tree, and as an oak,—whose substance is in them when they cast their leaves,—so the holy seed is in the midst thereof."

32. Only remember on what conditions. In the great Psalm of life, we are told that everything that a man döeth shall prosper, so only that he delight in the law of his God, that he hath not walked in the counsel of the wicked, nor sat in the seat of the scornful. Is it among these leaves of the perpetual Spring,—helpful leaves for the healing of the nations,—that we mean to have our part and place, or rather among the "brown skeletons of leaves that lag, the forest brook along"? For other leaves there are, and other streams that water them,—not water of life, but water of Acheron. Autumnal leaves there are that strew the brooks, in Vallombrosa. Remember you how the name of the place was changed: "Once called 'Sweet water' (Aqua bella), now, the Shadowy Vale." Portion in one or other name we must choose, all of us,—with the living olive, by the living fountains of waters, or with the wild fig trees, whose leafage of human soul is strewed along the brooks of death, in the eternal Vallombrosa.

CHAPTER IV.

THE FLOWER.

ROME, *Whit Monday*, 1874.

1. ON the quiet road leading from under the Palatine to the little church of St. Nereo and Achilleo, I met, yesterday morning, group after group of happy peasants heaped in pyramids on their triumphal carts, in Whit-Sunday dress, stout and clean, and gay in colour; and the women all with bright artificial roses in their hair, set with true natural taste, and well becoming them. This power of arranging wreath or crown of flowers for the head, remains to the people from classic times. And the thing that struck me most in the look of it was not so much the cheerfulness, as the dignity;—in a true sense, the *becomingness* and decorousness of the ornament. Among the ruins of the dead city, and the worse desolation of the work of its modern rebuilders, here was one element at least of honour, and order;—and, in these, of delight.

And these are the real significances of the flower itself. It is the utmost purification of the plant, and the utmost discipline. Where its tissue is blanched fairest, dyed purest, set in strictest rank, appointed to most chosen office, there—and created by the fact of this purity and function—is the flower.

2. But created, observe, by the purity and order, more than by the function. The flower exists for its own sake,—not for the fruit's sake. The production of the fruit is an added honour to it—is a granted consolation to us for its death. But the flower is the end of the seed,—not the seed of the flower. You are fond of cherries, perhaps; and think that the use of cherry blossom is to produce cherries. Not at all. The use of cherries is to produce cherry blossom; just as the use of bulbs is to produce hyacinths,—not of hyacinths to produce bulbs. Nay, that the flower can multiply by bulb, or root, or slip, as well as by seed, may show you at once how immaterial the seed-forming function is to the flower's existence. A flower is to the vegetable substance what a crystal is to the mineral. "Dust of sapphire," writes my friend Dr. John Brown to me, of the wood hyacinths of Scotland in the spring. Yes, that is so,—each bud more beautiful, itself, than perfectest jewel—*this*, indeed, jewel "of purest ray serene;" but, observe you, the glory is in the purity, the serenity, the radiance,—not in the mere continuance of the creature.

3. It is because of its beauty that its continuance

is worth Heaven's while. The glory of it is in being,—not in begetting; and in the spirit and substance,—not the change. For the earth also has its flesh and spirit. Every day of spring is the earth's Whit Sunday—Fire Sunday. The falling fire of the rainbow, with the order of its zones, and the gladness of its covenant,—you may eat of it, like Esdras; but you feed upon it only that you may see it. Do you think that flowers were born to nourish the blind?

Fasten well in your mind, then, the conception of order, and purity, as the essence of the flower's being, no less than of the crystal's. A ruby is not made bright to scatter round it child-rubies; nor a flower, but in collateral and added honour, to give birth to other flowers.

Two main facts, then, you have to study in every flower: the symmetry or order of it, and the perfection of its substance; first, the manner in which the leaves are placed for beauty of form; then the spinning and weaving and blanching of their tissue, for the reception of purest colour, or refining to richest surface.

4. First, the order: the proportion, and answering to each other, of the parts; for the study of which it becomes necessary to know what its parts are; and that a flower consists essentially of—— Well,

I really don't know what it consists essentially of. For some flowers have bracts, and stalks, and toruses, and calices, and corollas, and discs, and stamens, and pistils, and ever so many odds and ends of things besides, of no use at all, seemingly; and others have no bracts, and no stalks, and no toruses, and no calices, and no corollas, and nothing recognizable for stamens or pistils,—only, when they come to be reduced to this kind of poverty, one doesn't call them flowers; they get together in knots, and one calls them catkins, or the like, or forgets their existence altogether;—I haven't the least idea, for instance, myself, what an oak blossom is like; only I know its bracts get together and make a cup of themselves afterwards, which the Italians call, as they do the dome of St. Peter's, 'cupola'; and that it is a great pity, for their own sake as well as the world's, that they were not content with their ilex cupolas, which were made to hold something, but took to building these big ones upside-down, which hold nothing—*less* than nothing,—large extinguishers of the flame of Catholic religion. And for farther embarrassment, a flower not only is without essential consistence of a given number of parts, but it rarely consists, alone, of *itself*. One talks of a hyacinth as of a flower; but a hyacinth is any number of flowers. One does not talk of 'a heather'; when one says 'heath,' one means the whole plant, not the blossom,—because heath-bells, though they grow together for company's sake, do so in a voluntary sort of way, and are not fixed in their places; and yet, they depend on each other for effect, as much as a bunch of grapes.

5. And this grouping of flowers, more or less waywardly, is the most subtle part of their order, and the most difficult to represent. Take that cluster of bog-heather bells, for instance, Line-study I. You might think at first there were no lines in it worth study; but look at it more carefully. There are twelve bells in the cluster. There may be fewer, or more; but the bog-heath is apt to run into something near that number. They all grow together as close as they can, and on one side of the supporting branch only. The natural effect would be to bend the branch down; but the branch won't have that, and so leans back to carry them. Now you see the use of drawing the profile in the middle figure: it shows you the exactly balanced setting of the group,—not drooping, nor erect; but with a disposition to droop, tossed up by the leaning back of the stem. Then, growing as near as they can to each other, those in the middle get squeezed. Here is another quite

special character. Some flowers don't like being squeezed at all (fancy a squeezed convolvulus!); but these heather bells like it, and look all the prettier for it,—not the squeezed ones exactly, by themselves, but the cluster altogether, by their patience.

Then also the outside ones get pushed into a sort of star-shape, and in front show the colour of all their sides, and at the back the rich green cluster of sharp leaves that hold them; all this order being as essential to the plant as any of the more formal structures of the bell itself.

6. But the bog-heath has usually only one cluster of flowers to arrange on each branch. Take a spray of ling (Frontispiece), and you will find that the richest piece of Gothic spire-sculpture would be dull and graceless beside the grouping of the floral masses in their various life. But it is difficult to give the accuracy of attention necessary to see their beauty without drawing them; and still more difficult to draw them in any approximation to the truth before they change. This is indeed the fatallest obstacle to all good botanical work. Flowers, or leaves,—and especially the last,—can only be rightly drawn as they grow. And even then, in their loveliest spring action, they grow as you draw them, and will not stay quite the same creatures for half an hour.

7. I said in my inaugural lectures at Oxford, § 107, that real botany is not so much the description of plants as their biography. Without entering at all into the history of its fruitage, the life and death of the blossom *itself* is always an eventful romance, which must be completely told, if well. The grouping given to the various states of form between bud and flower is always the most important part of the design of the plant; and in the modes of its death are some of the most touching lessons, or symbolisms, connected with its existence. The utter loss and far-scattered ruin of the cistus and wild rose,—the dishonoured and dark contortion of the convolvulus,—the pale wasting of the crimson heath of Apennine, are strangely opposed by the quiet closing of the brown bells of the ling, each making of themselves a little cross as they die; and so enduring into the days of winter. I have drawn the faded beside the full branch, and know not which is the more beautiful.

8. This grouping, then, and way of treating each other in their gathered company, is the first and most subtle condition of form in flowers; and, observe, I don't mean, just now, the appointed and disciplined grouping, but the wayward and accidental. Don't confuse the beautiful consent of

the cluster in these sprays of heath with the legal strictness of a foxglove,—though that also has its divinity; but of another kind. That legal order of blossoming—for which we may wisely keep the accepted name, 'inflorescence,'—is itself quite a separate subject of study, which we cannot take up until we know the still more strict laws which are set over the flower itself.

9. I have in my hand a small red poppy which I gathered on Whit Sunday on the palace of the Cæsars. It is an intensely simple, intensely floral, flower. All silk and flame: a scarlet cup, perfect-edged all round, seen among the wild grass far away, like a burning coal fallen from Heaven's altars. You cannot have a more complete, a more stainless, type of flower absolute; inside and outside, *all* flower. No sparing of colour anywhere—no outside coarsenesses—no interior secrecies; open as the sunshine that creates it; fine-finished on both sides, down to the extremest point of insertion on its narrow stalk; and robed in the purple of the Cæsars.

Literally so. That poppy scarlet, so far as it could be painted by mortal hand, for mortal King, stays yet, against the sun, and wind, and rain, on the walls of the house of Augustus, a hundred yards from the spot where I gathered the weed of its desolation.

10. A pure *cup*, you remember it is; that much at least you cannot but remember, of poppy-form among the cornfields; and it is best, in beginning, to think of every flower as essentially a cup. There are flat ones, but you will find that most of these are really groups of flowers, not single blossoms; and there are out-of-the-way and quaint ones, very difficult to define as of any shape; but even these have a cup to begin with, deep down in them. You had better take the idea of a cup or vase, as the first, simplest, and most general form of true flower.

The botanists call it a corolla, which means a garland, or a kind of crown; and the word is a very good one, because it indicates that the flower-cup is made, as our clay cups are, on a potter's wheel; that it is essentially a *revolute* form—a whirl or (botanically) 'whorl' of leaves; in reality successive round the base of the urn they form.

11. Perhaps, however, you think poppies in general are not much like cups. But the flower in my hand is a—poverty-*stricken* poppy, I was going to write, ——poverty-*strengthened* poppy, I mean. On richer ground, it would have gushed into flaunting breadth of untenable purple—flapped its inconsistent scarlet vaguely to the wind—dropped the pride of its petals over my hand in an hour after I gathered

it. But this little rough-bred thing, a Campagna pony of a poppy, is as bright and strong to-day as yesterday. So that I can see exactly where the leaves join or lap over each other; and when I look down into the cup, find it to be composed of four leaves altogether,—two smaller, set within two larger.

12. Thus far (and somewhat farther) I had written in Rome; but now, putting my work together in Oxford, a sudden doubt troubles me, whether all poppies have two petals smaller than the other two. Whereupon I take down an excellent little school-book on botany—the best I've yet found, thinking to be told quickly; and I find a great deal about opium; and, apropos of opium, that the juice of common celandine is of a bright orange colour; and I pause for a bewildered five minutes, wondering if a celandine is a poppy, and how many petals *it* has: going on again—because I must, without making up my mind, on either question—I am told to "observe the floral receptacle of the Californian genus Eschscholtzia." Now I can't observe anything of the sort, and I don't want to; and I wish California and all that's in it were at the deepest bottom of the Pacific. Next I am told to compare the poppy and waterlily; and I can't do that, neither—though I should like to; and there's the end of the article;

and it never tells me whether one pair of petals is always smaller than the other, or not. Only I see it says the corolla has four petals. Perhaps a celandine may be a double poppy, and have eight. I know they're tiresome irregular things, and I mustn't be stopped by them ;*—at any rate, my Roman poppy knew what it was about, and had its two couples of leaves in clear subordination, of which at the time I went ón to inquire farther, as follows.

13. The next point is, what shape are the petals of? And that is easier asked than answered; for when you pull them off, you find they won't lie flat, by any means, but are each of them cups, or rather shells, themselves; and that it requires as much conchology as would describe a cockle, before you can properly give account of a single poppy leaf. Or of a single *any* leaf—for all leaves are either shells, or boats, (or solid, if not hollow, masses,) and cannot be represented in flat outline. But, laying these as flat as they will lie on a sheet of paper, you will find the piece they hide of the paper they lie on can be drawn; giving

* Just in time, finding a heap of gold under an oak tree some thousand years old, near Arundel, I've made them out: Eight, divided by three; that is to say, three couples of petals, with two odd little ones inserted for form's sake. No wonder I couldn't decipher them by memory.

approximately the shape of the outer leaf as at A, that of the inner as at B, Fig. 4; which you will find very difficult lines to draw, for they are each composed of two curves, joined, as in Fig. 5; all above the line *a b* being the outer edge of the leaf, but joined so subtly to the side that the least break in drawing the line spoils the form.

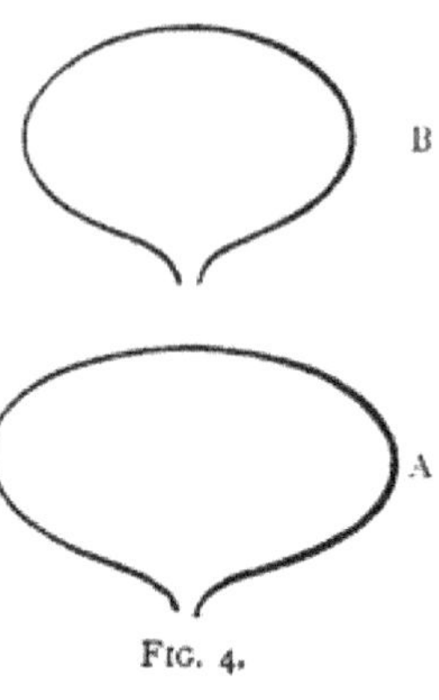

FIG. 4.

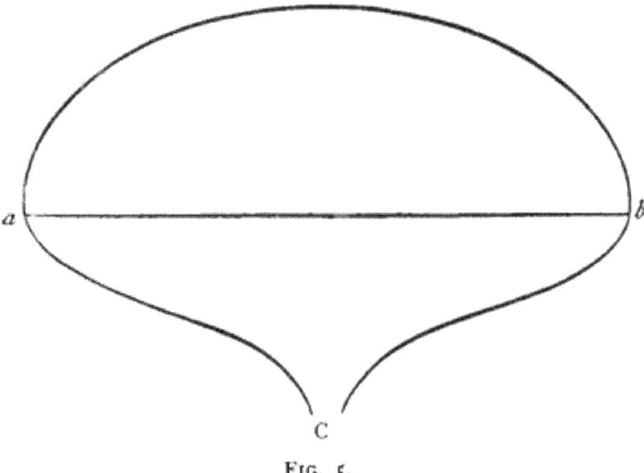

FIG. 5.

14. Now every flower petal consists essentially of these two parts, variously proportioned and outlined.

It expands from C to *a b*; and closes in the external line, and for this reason.

Considering every flower under the type of a cup, the first part of the petal is that in which it expands from the bottom to the rim; the second part, that in which it terminates itself on reaching the rim. Thus let the three circles, A B C, Fig. 6, represent

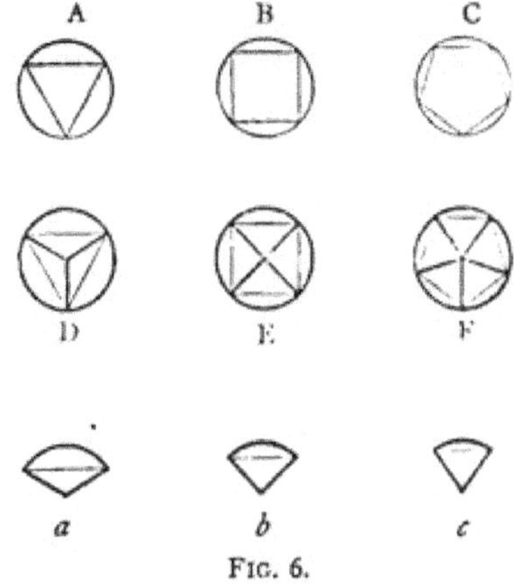

Fig. 6.

the undivided cups of the three great geometrical orders of flowers—trefoil, quatrefoil, and cinquefoil.

Draw in the first an equilateral triangle, in the second a square, in the third a pentagon; draw the dark lines from centres to angles; (D E F): then (*a*) the third part of D; (*b*) the fourth part of E, (*c*) the fifth part of F, are the normal outline forms of the petals of the three families; the relations between the developing angle and limiting curve being varied according to the depth of cup, and the degree of connection between the petals. Thus a rose folds them over one another, in the bud; a convolvulus twists them,—the one expanding into a flat cinquefoil of separate petals, and the other into a deep-welled cinquefoil of connected ones.

I find an excellent illustration in Veronica Polita, one of the most perfectly graceful of field plants because of the light alternate flower stalks, each with its leaf at the base; the flower itself a quatrefoil, of which the largest and least petals are uppermost. Pull one off its calyx (draw, if you can, the outline of the striped blue upper petal with the jagged edge of pale gold below), and then examine the relative shapes of the lateral, and least upper petal. Their under surface is very curious, as if covered with white paint; the blue stripes above, in the direction of their growth, deepening the more delicate colour with exquisite insistence.

A lilac blossom will give you a pretty example of the expansion of the petals of a quatrefoil above the edge of the cup or tube; but I must get back to our poppy at present.

15. What outline its petals really have, however, is little shown in their crumpled fluttering; but that very crumpling arises from a fine floral cha-

racter which we do not enough value in them. We usually think of the poppy as a coarse flower; but it is the most transparent and delicate of all the blossoms of the field. The rest—nearly all of them—depend on the *texture* of their surfaces for colour. But the poppy is painted *glass;* it never glows so brightly as when the sun shines through it. Wherever it is seen—against the light or with the light—always, it is a flame, and warms the wind like a blown ruby.

In these two qualities, the accurately balanced form, and the perfectly infused colour of the petals, you have, as I said, the central being of the flower. All the other parts of it are necessary, but we must follow them out in order.

16. Looking down into the cup, you see the green boss divided by a black star,—of six rays only,—and surrounded by a few black spots. My rough-nurtured poppy contents itself with these for its centre; a rich one would have had the green boss divided by a dozen of rays, and surrounded by a dark crowd of crested threads.

This green boss is called by botanists the pistil, which word consists of the two first syllables of the Latin pistillum, otherwise more familiarly Englished into 'pestle.' The meaning of the botanical word is of course, also, that the central part of a flower-cup has to it something of the relations that a pestle has to a mortar! Practically, however, as this pestle has no pounding functions, I think the word is misleading as well as ungraceful; and that we may find a better one after looking a little closer into the matter. For this pestle is divided generally into three very distinct parts: there is a storehouse at the bottom of it for the seeds of the plant; above this, a shaft, often of considerable length in deep cups, rising to the level of their upper edge, or above it; and at the top of these shafts an expanded crest. This shaft the botanists call 'style,' from the Greek word for a pillar; and the crest of it—I do not know why—stigma, from the Greek word for 'spot.' The storehouse for the seeds they call the 'ovary,' from the Latin ovum, an egg. So you have two-thirds of a Latin word, (pistil)—awkwardly and disagreeably edged in between pestle and pistol—for the whole thing; you have an English-Latin word (ovary) for the bottom of it; an English-Greek word (style) for the middle; and a pure Greek word (stigma) for the top.

17. This is a great mess of language, and all the worse that the words style and stigma have both of them quite different senses in ordinary and scholarly English from this forced botanical one. And I will venture therefore, for my own pupils, to

put the four names altogether into English. Instead of calling the whole thing a pistil, I shall simply call it the pillar. Instead of 'ovary,' I shall say 'Treasury' (for a seed isn't an egg, but it *is* a treasure). The style I shall call the 'Shaft,' and the stigma the 'Volute.' So you will have your entire pillar divided into the treasury, at its base, the shaft, and the volute; and I think you will find these divisions easily remembered, and not unfitted to the sense of the words in their ordinary use.

18. Round this central, but, in the poppy, very stumpy, pillar, you find a cluster of dark threads, with dusty pendants or cups at their ends. For these the botanists' name 'stamens,' may be conveniently retained, each consisting of a 'filament,' or thread, and an 'anther,' or blossoming part.

And in this rich corolla, and pillar, or pillars, with their treasuries, and surrounding crowd of stamens, the essential flower consists. Fewer than these several parts, it cannot have, to be a flower at all; of these, the corolla leads, and is the object of final purpose. The stamens and the treasuries are only there in order to produce future corollas, though often themselves decorative in the highest degree.

These, I repeat, are all the essential parts of a flower. But it would have been difficult, with any other than the poppy, to have shown you them alone; for nearly all other flowers keep with them, all their lives, their nurse or tutor leaves,—the group which, in stronger and humbler temper, protected them in their first weakness, and formed them to the first laws of their being. But the poppy casts these tutorial leaves away. It is the finished picture of impatient and luxury-loving youth,—at first too severely restrained, then casting all restraint away, —yet retaining to the end of life unseemly and illiberal signs of its once compelled submission to laws which were only pain,—not instruction.

19. Gather a green poppy bud, just when it shows the scarlet line at its side; break it open and unpack the poppy. The whole flower is there complete in size and colour,—its stamens full-grown, but all packed so closely that the fine silk of the petals is crushed into a million of shapeless wrinkles. When the flower opens, it seems a deliverance from torture: the two imprisoning green leaves are shaken to the ground; the aggrieved corolla smooths itself in the sun, and comforts itself as it can; but remains visibly crushed and hurt to the end of its days.

20. Not so flowers of gracious breeding. Look at these four stages in the young life of a primrose, Fig. 7. First confined, as strictly as the poppy within five pinching green leaves, whose points

close over it, the little thing is content to remain a child, and finds its nursery large enough. The green leaves unclose their points,—the little yellow ones

Fig. 7.

peep out, like ducklings. They find the light delicious, and open wide to it; and grow, and grow, and throw themselves wider at last into their perfect rose. But they never leave their old nursery for all that; it and they live on together; and the nursery seems a part of the flower.

21. Which is so, indeed, in all the loveliest flowers; and, in usual botanical parlance, a flower is said to consist of its calyx, (or *hiding* part—Calypso having rule over it,) and corolla, or garland part, Proserpina having rule over it. But it is better to think of them always as

separate; for this calyx, very justly so named from its main function of concealing the flower, in its youth is usually green, not coloured, and shows its separate nature by pausing, or at least greatly lingering, in its growth, and modifying itself very slightly, while the corolla is forming itself through active change. Look at the two, for instance, through the youth of a pease blossom, Fig. 8.

Fig. 8.

The entire cluster at first appears pendent in this manner, the stalk bending round on purpose to put it into that position. On which all the little buds, thinking themselves ill-treated, determine not to submit to anything of the sort, turn their points upwards persistently, and determine that—at any cost of trouble—they will get nearer the sun. Then they begin to open, and let out their corollas. I give the progress of one only (Fig. 9).* It chances to be engraved the reverse way from the bud; but that is of no consequence.

At first, you see the long lower point of the

* Figs. 8 and 9 are both drawn and engraved by Mr. Burgess.

calyx thought that *it* was going to be the head of the family, and curls upwards eagerly. Then the little corolla steals out; and soon does away with that impression on the mind of the calyx. The corolla soars up with widening wings, the abashed calyx retreats beneath; and finally the great upper leaf of corolla—not pleased at having its back still turned to the light, and its face down—throws itself entirely back, to look at the sky, and nothing else;—and your blossom is complete.

Fig. 9.

Keeping, therefore, the ideas of calyx and corolla entirely distinct, this one general point you may note of both: that, as a calyx is originally folded tight over the flower, and has to open deeply to let it out, it is nearly always composed of sharp pointed leaves like the segments of a balloon; while corollas, having to open out as wide as possible to show themselves, are typically like cups or plates, only cut into their edges here and there, for ornamentation's sake.

22. And, finally, though the corolla is essentially the floral group of leaves, and usually receives the glory of colour for itself only, this glory and delight may be given to any other part of the group; and, as if to show us that there is no really dishonoured or degraded membership, the stalks and leaves in some plants, near the blossom, flush in sympathy with it, and become themselves a part of the effectively visible flower;—Eryngo—Jura hyacinth, (comosus,) and the edges of upper stems and leaves in many plants; while others, (Geranium lucidum,) are made to delight us with their leaves rather than their blossoms; only I suppose, in these, the scarlet leaf colour is a kind of early autumnal glow,—a beautiful hectic, and foretaste, in sacred youth, of sacred death.

I observe, among the speculations of modern science, several, lately, not uningenious, and highly industrious, on the subject of the relation of colour in flowers, to insects—to selective development, etc., etc. There *are* such relations, of course. So also, the blush of a girl, when she first perceives the faltering in her lover's step as he draws near, is related essentially to the existing state of her stomach; and to the state of it through all the years of her previous existence. Nevertheless, neither love, chastity, nor blushing, are merely exponents of digestion.

All these materialisms, in their unclean stupidity, are essentially the work of human bats; men of semi-faculty or semi-education, who are more or less incapable of so much as seeing, much less thinking about, colour; among whom, for one-sided intensity, even Mr. Darwin must be often ranked, as in his vespertilian treatise on the ocelli of the Argus pheasant, which he imagines to be artistically gradated, and perfectly imitative of a ball and socket. If I had him here in Oxford for a week, and could force him to try to copy a feather by Bewick, or to draw for himself a boy's thumbed marble, his notions of feathers, and balls, would be changed for all the rest of his life. But his ignorance of good art is no excuse for the acutely illogical simplicity of the rest of his talk of colour in the "Descent of Man." Peacocks' tails, he thinks, are the result of the admiration of blue tails in the minds of well-bred peahens,—and similarly, mandrills' noses the result of the admiration of blue noses in well-bred baboons. But it never occurs to him to ask why the admiration of blue noses is healthy in baboons, so that it develops their race properly, while similar maidenly admiration either of blue noses or red noses in men would be improper, and develop the race improperly. The word itself 'proper' being one of which he has never asked, or guessed, the meaning. And when he imagined the gradation of the cloudings in feathers to represent successive generation, it never occurred to him to look at the much finer cloudy gradations in the clouds of dawn themselves; and explain the modes of sexual preference and selective development which had brought *them* to their scarlet glory, before the cock could crow thrice.

Putting all these vespertilian speculations out of our way, the human facts concerning colour are briefly these. Wherever men are noble, they love bright colour; and wherever they can live healthily, bright colour is given them—in sky, sea, flowers, and living creatures.

On the other hand, wherever men are ignoble and sensual, they endure without pain, and at last even come to like, (especially if artists,) mud-colour and black, and to dislike rose-colour and white. And wherever it is unhealthy for them to live, the poisonousness of the place is marked by some ghastly colour in air, earth, or flowers.

There are, of course, exceptions to all such widely founded laws; there are poisonous berries of scarlet, and pestilent skies that are fair. But, if we once honestly compare a venomous wood-fungus, rotting into black dissolution of dripped slime at its edges, with a spring gentian; or a puff adder with a salmon trout, or a fog in Bermondsey with a clear sky at Berne, we shall get hold of the entire question on its right side; and be able afterwards to study at our leisure, or accept without doubt or trouble, facts of apparently contrary meaning. And the practical lesson which I wish to leave with the reader is, that lovely flowers, and green trees growing in the open air, are the proper guides of men to the places which their Maker intended them to inhabit; while the flowerless and treeless deserts—of reed, or sand, or rock,—are meant to be either heroically invaded and redeemed, or surrendered to the wild creatures which are appointed for them; happy and wonderful in their wild abodes.

Nor is the world so small but that we may yet leave in it also unconquered spaces of beautiful solitude; where the chamois and red deer may wander fearless,—nor any fire of avarice scorch from the Highlands of Alp, or Grampian, the rapture of the heath, and the rose.

CHAPTER V.

PAPAVER RHOEAS.

BRANTWOOD, *July 11th*, 1875.

1. CHANCING to take up yesterday a favourite old book, Mavor's British Tourists, (London, 1798,) I found in its fourth volume a delightful diary of a journal made in 1782 through various parts of England, by Charles P. Moritz of Berlin.

And in the fourteenth page of this diary I find the following passage, pleasantly complimentary to England :—

"The slices of bread and butter which they give you with your tea are as thin as poppy leaves. But there is another kind of bread and butter usually eaten with tea, which is toasted by the fire, and is incomparably good. This is called 'toast.'"

I wonder how many people, nowadays, whose bread and butter was cut too thin for them, would think of comparing the slices to poppy leaves? But this was in the old days of travelling, when people did not whirl themselves past corn-fields, that they might have more time to walk on paving-stones;

and understood that poppies did not mingle their scarlet among the gold, without some purpose of the poppy-Maker that they should be looked at.

Nevertheless, with respect to the good and polite German's poetically-contemplated, and finely æsthetic, tea, may it not be asked whether poppy leaves themselves, like the bread and butter, are not, if we may venture an opinion—*too* thin,—im-*properly* thin? In the last chapter, my reader was, I hope, a little anxious to know what I meant by saying that modern philosophers did not know the meaning of the word 'proper,' and may wish to know what I mean by it myself. And this I think it needful to explain before going farther.

2. In our English prayer-book translation, the first verse of the ninety-third Psalm runs thus: "The Lord is King; and hath put on glorious apparel." And although, in the future republican world, there are to be no lords, no kings, and no glorious apparel, it will be found convenient, for botanical purposes, to remember what such things once were; for when I said of the poppy, in last chapter, that it was "robed in the purple of the Cæsars," the words gave, to any one who had a clear idea of a Cæsar, and of his dress, a better, and even *stricter*, account of the flower than if I had only said, with Mr. Sowerby, "petals bright scarlet;" which might just as well

have been said of a pimpernel, or scarlet geranium; —but of neither of these latter should I have said "robed in purple of Cæsars." What I meant was, first, that the poppy leaf looks dyed through and through, like glass, or Tyrian tissue; and not merely painted: secondly, that the splendour of it is proud,—almost insolently so. Augustus, in his glory, might have been clothed like one of these; and Saul; but not David, nor Solomon; still less the teacher of Solomon, when He puts on 'glorious apparel.'

3. Let us look, however, at the two translations of the same verse.

In the Vulgate it is "Dominus regnavit; decorem indutus est;" He has put on 'becomingness,'—decent apparel, rather than glorious.

In the Septuagint it is ευπρεπειὰ—*well*-becomingness; an expression which, if the reader considers, must imply certainly the existence of an opposite idea of possible '*ill*-becomingness,'—of an apparel which should, in just as accurate a sense, belong appropriately to the creature invested with it, and yet not be glorious, but inglorious, and not well-becoming, but ill-becoming. The mandrill's blue nose, for instance, already referred to,—can we rightly speak of this as 'ευπρεπειὰ'? Or the stings, and minute, colourless blossoming of the nettle?

May we call these a glorious apparel, as we may the glowing of an alpine rose?

You will find on reflection, and find more convincingly the more accurately you reflect, that there is an absolute sense attached to such words as 'decent,' 'honourable,' 'glorious,' or 'καλος,' contrary to another absolute sense in the words 'indecent,' 'shameful,' 'vile,' or 'αἰσχρος.'

And that there is every degree of these absolute qualities visible in living creatures; and that the divinity of the Mind of man is in its essential discernment of what is καλον from what is αἰσχρον, and in his preference of the kind of creatures which are decent, to those which are indecent; and of the kinds of thoughts, in himself, which are noble, to those which are vile.

4. When therefore I said that Mr. Darwin, and his school,* had no conception of the real meaning of the word 'proper,' I meant that they conceived the qualities of things only as their 'properties,' but not as their 'becomingnesses;' and seeing that dirt is proper to a swine, malice to a monkey, poison to a nettle, and folly to a fool, they called a nettle *but* a nettle, and the faults of fools but folly; and never saw the difference between ugliness and beauty

* Of Vespertilian science generally, compare 'Eagle's Nest,' pp. 25, and 179 (of the "Revised Series," and pp. 28, 206-7 of the small edition)

absolute, decency and indecency absolute, glory or shame absolute, and folly or sense absolute.

Whereas, the perception of beauty, and the power of defining physical character, are based on moral instinct, and on the power of defining animal or human character. Nor is it possible to say that one flower is more highly developed, or one animal of a higher order, than another, without the assumption of a divine law of perfection to which the one more conforms than the other.

5. Thus, for instance. That it should ever have been an open question with me whether a poppy had always two of its petals less than the other two, depended wholly on the hurry and imperfection with which the poppy carries out its plan. It never would have occurred to me to doubt whether an iris had three of its leaves smaller than the other three, because an iris always completes itself to its own ideal. Nevertheless, on examining various poppies, as I walked, this summer, up and down the hills between Sheffield and Wakefield, I find the subordination of the upper and lower petals entirely necessary and normal; and that the result of it is to give two distinct profiles to the poppy cup, the difference between which, however, we shall see better in the yellow Welsh poppy, at present called Meconopsis Cambrica, but which, in the Oxford schools, will be 'Papaver

cruciforme'—'Crosslet Poppy,'—first, because all our botanical names must be in Latin if possible;

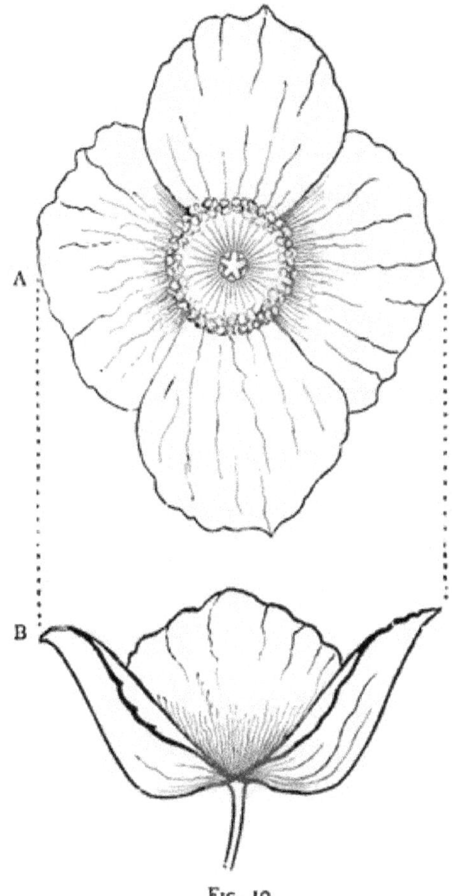

Fig. 10.

Greek only allowed when we can do no better;

secondly, because meconopsis is barbarous Greek; thirdly, and chiefly, because it is little matter whether this poppy be Welsh or English; but very needful that we should observe, wherever it grows, that the petals are arranged in what used to be, in my young days, called a diamond shape,* as at A, Fig. 10, the two narrow inner ones at right angles to, and projecting farther than, the two outside broad ones; and that the two broad ones, when the flower is seen in profile, as at B, show their margins folded back, as indicated by the thicker lines, and have a profile curve, which is only the softening, or melting away into each other, of two straight lines. Indeed, when the flower is younger, and quite strong, both its profiles, A and B, Fig. 11, are nearly straight-sided; and

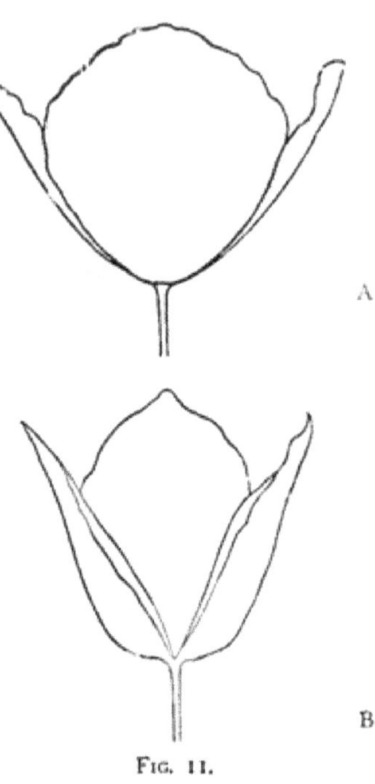

Fig. 11.

* The mathematical term is 'rhomb.'

always, be it young or old, one broader than the other, so as to give the flower, seen from above, the shape of a contracted cross, or crosslet.

6. Now I find no notice of this flower in Gerarde; and in Sowerby, out of eighteen lines of closely printed descriptive text, no notice of its crosslet form, while the petals are only stated to be "roundish-concave," terms equally applicable to at least one-half of all flower petals in the world. The leaves are *said* to be very deeply pinnately partite; but *drawn*—as neither pinnate nor partite!

And this is your modern cheap science, in ten volumes. Now I haven't a quiet moment to spare for drawing this morning; but I merely give the main relations of the petals, A, and blot in the wrinkles of one of the lower ones, B, Fig. 12; and yet in this rude sketch you will feel, I believe, there is something specific which could not belong to any other flower. But all proper description is impossible without careful profiles of each petal laterally and across it. Which I may not find time to draw for any poppy whatever, because they none of them have well-becomingness enough to make it worth my while, being all more or less weedy, and ungracious, and mingled of good and evil. Whereupon rises before me, ghostly and untenable, the general question, 'What is a weed?' and, impatient for answer, the particular question, 'What is a poppy?' I choose, for instance, to call this yellow flower a poppy, instead of a "likeness to poppy," which the botanists meant to call it, in their bad Greek. I choose also to call a poppy, what the botanists have called "glaucous thing," (glaucium). But where and when shall I stop calling things poppies? This is certainly a question to be settled at once, with others appertaining to it.

A

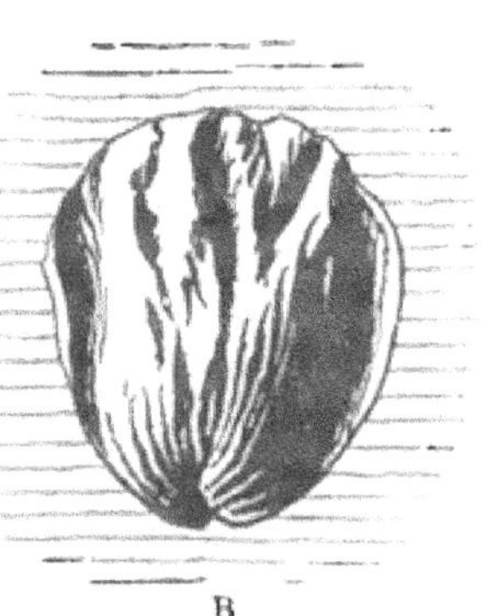

B

Fig. 12.

7. In the first place, then, I mean to call every flower either one thing or another, and not an 'aceous' thing, only half something or half another. I mean to call this plant now in my hand, either a poppy or not a poppy; but not poppaceous. And this other, either a thistle or not a thistle; but not thistlaceous. And this other, either a nettle or not a nettle; but not nettlaceous. I know it will be

very difficult to carry out this principle when tribes of plants are much extended and varied in type: I shall persist in it, however, as far as possible; and when plants change so much that one cannot with any conscience call them by their family name any more, I shall put them aside somewhere among families of poor relations, not to be minded for the present, until we are well acquainted with the better bred circles. I don't know, for instance, whether I shall call the Burnet 'Grass-rose,' or put it out of court for having no petals; but it certainly shall not be called rosaceous; and my first point will be to make sure of my pupils having a clear idea of the central and unquestionable forms of thistle, grass, or rose, and assigning to them pure Latin, and pretty English, names,—classical, if possible; and at least intelligible and decorous.

8. I return to our present special question, then, What is a poppy? and return also to a book I gave away long ago, and have just begged back again, Dr. Lindley's 'Ladies' Botany.' For without at all looking upon ladies as inferior beings, I dimly hope that what Dr. Lindley considers likely to be intelligible to *them*, may be also clear to their very humble servant.

The poppies, I find, (page 19, vol. i.) differ from crowfeet in being of a stupifying instead of a burning nature, and in generally having two sepals and twice two petals; "but as some poppies have three sepals, and twice three petals, the number of these parts is not sufficiently constant to form an essential mark." Yes, I know that, for I found a superb six-petaled poppy, spotted like a cistus, the other day in a friend's garden. But then, what makes it a poppy still? That it is of a stupifying nature, and itself so stupid that it does not know how many petals it should have, is surely not enough distinction?

9. Returning to Lindley, and working the matter farther out with his help, I think this definition might stand. "A poppy is a flower which has either four or six petals, and two or more treasuries, united into one; containing a milky, stupifying fluid in its stalks and leaves, and always throwing away its calyx when it blossoms."

And indeed, every flower which unites all these characters, we shall, in the Oxford schools, call 'poppy,' and 'Papaver;' but when I get fairly into work, I hope to fix my definitions into more strict terms. For I wish all my pupils to form the habit of asking, of every plant, these following four questions, in order, corresponding to the subject of these opening chapters, namely, "What root has it? what leaf? what flower? and what stem?" And, in this definition of poppies, nothing whatever is said about the root; and not only I don't know myself what a poppy root is like, but in all Sowerby's poppy section, I find no word whatever about that matter.

10. Leaving, however, for the present, the root unthought of, and contenting myself with Dr. Lindley's characteristics, I shall place, at the head of the whole group, our common European wild poppy, Papaver Rhoeas, and, with this, arrange the nine following other flowers thus,—opposite.

I must be content at present with determining the Latin names for the Oxford schools; the English ones I shall give as they chance to occur to me, in Gerarde and the classical poets who wrote before the English revolution. When no satisfactory name is to be found, I must try to invent one; as, for instance, just now, I don't like Gerarde's 'Corn-rose' for Papaver Rhoeas, and must coin another; but this can't be done by thinking; it will come into my head some day, by chance. I might try at it straightforwardly for a week together, and not do it.

The Latin names must be fixed at once, somehow; and therefore I do the best I can, keeping as much respect for the old nomenclature as possible, though this involves the illogical practice of giving the epithet sometimes from the flower, (violaceum,

NAME IN OXFORD CATALOGUE.	DIOSCORIDES.	IN PRESENT BOTANY.
1. Papaver Rhoeas	μηκων ῥοιας	Papaver Rhoeas
2. P. Hortense	μ. κηπευτη*	P. Hortense
3. P. Elatum	μ. θυλακίτις†	P. Lamottei
4. P. Argemone	. . .	P. Argemone
5. P. Echinosum	. . .	P. Hybridum
6. P. Violaceum	. .	Roemeria Hybrida
7. P. Cruciforme	. . .	Meconopsis Cambrica
8. P. Corniculatum	μ. κερατίτις	Glaucium Corniculatum
9. P. Littorale	μ. παραλιος	Glaucium Luteum
10. P. Chelidonium	. .	Chelidonium Majus

* ἧς τὸ σπέρμα ἀρτοποιεῖται.

† ἐπίμηκες ἔχουσα τὸ κεφάλιον. Dioscorides makes no effort to distinguish species, but gives the different names as if merely used in different places.

cruciforme), and sometimes from the seed vessel, (elatum, echinosum, corniculatum). Guarding this distinction, however, we may perhaps be content to call the six last of the group, in English, Urchin Poppy, Violet Poppy, Crosslet Poppy, Horned Poppy, Beach Poppy, and Welcome Poppy. I don't think the last flower pretty enough to be connected more directly with the swallow, in its English name.

11. I shall be well content if my pupils know these ten poppies rightly; all of them at present wild in our own country, and, I believe, also European in range: the head and type of all being the common wild poppy of our corn-fields for which the name 'Papaver Rhoeas,' given it by Dioscorides, Gerarde, and Linnæus, is entirely authoritative, and we will therefore at once examine the meaning, and reason, of that name.

12. Dioscorides says the name belongs to it "διὰ τὸ ταχέως τὸ ἄνθος ἀποβάλλειν," "because it casts off its bloom quickly," from ῥέω, (rheo) in the sense of shedding.* And this indeed it does,—first calyx, then corolla;—you may translate it 'swiftly ruinous' poppy, but notice, in connection with this idea, how it droops its head *before* blooming; an action which,

* It is also used sometimes of the garden poppy, says Dioscorides, "διὰ τὸ ῥεῖν ἐξ αὐτῆς τὸν ὀπόν"—"because the sap, opium, flows from it."

I doubt not, mingled in Homer's thought with the image of its depression when filled by rain, in the passage of the Iliad, which, as I have relieved your memory of three unnecessary names of poppy families, you have memory to spare for learning.

> "μήκων δ' ὣς ἑτέρωσε κάρη βάλεν, ἥτ' ἐνὶ κήπῳ
> καρπῷ βριθομένη, νοτίῃσί τε εἰαρινῇσιν
> ὣς ἑτέρωσ' ἤμυσε κάρη πήληκι βαρυνθέν."

"And as a poppy lets its head fall aside, which in a garden is loaded with its fruit, and with the soft rains of spring, so the youth drooped his head on one side; burdened with the helmet."

And now you shall compare the translations of this passage, with its context, by Chapman and Pope—(or the school of Pope), the one being by a man of pure English temper, and able therefore to understand pure Greek temper; the other infected with all the faults of the falsely classical school of the Renaissance.

First I take Chapman:—

> "His shaft smit fair Gorgythion, of Priam's princely race,
> Who in Æpina was brought forth, a famous town in Thrace,
> By Castianeira, that for form was like celestial breed.
> And as a crimson poppy-flower, surcharged with his seed,
> And vernal humours falling thick, declines his heavy brow,
> So, a-oneside, his helmet's weight his fainting head did bow."

Next, Pope :—

"He missed the mark ; but pierced Gorgythio's heart,
And drenched in royal blood the thirsty dart :
(Fair Castianeira, nymph of form divine,
This offspring added to King Priam's line).
As full-blown poppies, overcharged with rain,
Decline the head, and drooping kiss the plain,
So sinks the youth : his beauteous head, depressed
Beneath his helmet, drops upon his breast."

13. I give you the two passages in full, trusting that you may so feel the becomingness of the one, and the gracelessness of the other. But note farther, in the Homeric passage, one subtlety which cannot enough be marked even in Chapman's English, that his second word, ἤμυσε, is employed by him both of the stooping of ears of corn, under wind, and of Troy stooping to its ruin ; * and otherwise, in good Greek writers, the word is marked as having such specific sense of men's drooping under weight ; or towards death, under the burden of fortune which they have no more strength to sustain ; † compare the passage I quoted from

* See all the passages quoted by Liddell.

† I find this chapter rather tiresome on re-reading it myself, and cancel some farther criticism of the imitation of this passage by Virgil, one of the few pieces of the Æneid which are purely and vulgarly imitative, rendered also false as well as weak by the introducing sentence, "Volvitur Euryalus leto," after which the simile of the drooping flower is absurd. Of criticism, the chief use of which is to warn all sensible men from such business, the following abstract of Diderot's notes on the passage, given in the 'Saturday Review' for April 29, 1871, is

Plato, ('Crown of Wild Olive,' p. 95 of the "Revised Series," and p. 111 of the small edition): "And bore lightly the burden of gold and of possessions."

worth preserving. (Was the French critic really not aware that Homer *had* written the lines his own way?)

"Diderot illustrates his theory of poetical hieroglyphs by no quotations, but we can show the manner of his minute and sometimes fanciful criticism by repeating his analysis of the passage of Virgil wherein the death of Euryalus is described:—

'Pulchrosque per artus
It cruor, inque humeros cervix collapsa recumbit;
Purpureus veluti cum flos succisus aratro
Languescit moriens; lassove papavera collo
Demisere caput, pluvia cum forte gravantur.'

"The sound of 'It cruor,' according to Diderot, suggests the image of a jet of blood; 'cervix collapsa recumbit,' the fall of a dying man's head upon his shoulder; 'succisus' imitates the use of a cutting scythe (not plough); 'demisere' is as soft as the eye of a flower; 'gravantur,' on the other hand, has all the weight of a calyx, filled with rain; 'collapsa' marks an effort and a fall, and similar double duty is performed by 'papavera,' the first two syllables symbolizing the poppy upright, the last two the poppy bent. While thus pursuing his minute investigations, Diderot can scarcely help laughing at himself, and candidly owns that he is open to the suspicion of discovering in the poem beauties which have no existence. He therefore qualifies his eulogy by pointing out two faults in the passage. 'Gravantur,' notwithstanding the praise it has received, is a little too heavy for the light head of a poppy, even when filled with water. As for 'aratro,' coming as it does after the hiss of 'succisus,' it is altogether abominable. Had Homer written the lines, he would have ended with some hieroglyph, which would have continued the hiss or described the fall of a flower. To the hiss of 'succisus' Diderot is warmly attached. Not by mistake, but in order to justify the sound, he ventures to translate 'aratrum' into 'scythe,' boldly and rightly declaring in a marginal note that this is not the meaning of the word."

And thus you will begin to understand how the poppy became in the heathen mind the type at once of power, or pride, and of its loss; and therefore, both while Virgil represents the white nymph Nais, "pallentes violas, et summa papavera carpens,"—gathering the pale flags, and the highest poppies,—and the reason for the choice of this rather than any other flower, in the story of Tarquin's message to his son.

14. But you are next to remember the word Rhoeas in another sense. Whether originally intended or afterwards caught at, the resemblance of the word to 'Rhoea,' a pomegranate, mentally connects itself with the resemblance of the poppy head to the pomegranate fruit.

And if I allow this flower to be the first we take up for careful study in Proserpina, on account of its simplicity of form and splendour of colour, I wish you also to remember, in connection with it, the cause of Proserpine's eternal captivity—her having tasted a pomegranate seed,—the pomegranate being in Greek mythology what the apple is in the Mosaic legend; and, in the whole worship of Demeter, associated with the poppy by a multitude of ideas which are not definitely expressed, but can only be gathered out of Greek art and literature, as we learn their symbolism. The chief character on which these thoughts are founded is the fulness of seed in the poppy and pomegranate, as an image of life; then the forms of both became adopted for beads or bosses in ornamental art; the pomegranate remains more distinctly a Jewish and Christian type, from its use in the border of Aaron's robe, down to the fruit in the hand of Angelico's and Botticelli's Infant Christs; while the poppy is gradually confused by the Byzantine Greeks with grapes; and both of these with palm fruit. The palm, in the shorthand of their art, gradually becomes a symmetrical branched ornament with two pendent bosses; this is again confused with the Greek iris, (Homer's blue iris, and Pindar's water-flag,)—and the Florentines, in adopting Byzantine ornament, read it into their own Fleur-de-lys; but insert two poppy heads on each side of the entire foil, in their finest heraldry.

15. Meantime the definitely intended poppy, in late Christian Greek art of the twelfth century, modifies the form of the Acanthus leaf with its own, until the northern twelfth century workman takes the thistle-head for the poppy, and the thistle-leaf for acanthus. The true poppy-head remains in the south, but gets more and more confused with grapes, till the Renaissance carvers are content with any kind of boss full of seed, but insist on such boss

or bursting globe as some essential part of their ornament;—the bean-pod for the same reason (not without Pythagorean notions, and some of republican election) is used by Brunelleschi for main decoration of the lantern of Florence Duomo; and, finally, the ornamentation gets so shapeless, that M. Viollet-le-Duc, in his 'Dictionary of Ornament,' loses trace of its origin altogether, and fancies the later forms were derived from the spadix of the arum.

16. I have no time to enter into farther details; but through all this vast range of art, note this singular fact, that the wheat-ear, the vine, the fleur-de-lys, the poppy, and the jagged leaf of the acanthus-weed, or thistle, occupy the entire thoughts of the decorative workmen trained in classic schools, to the exclusion of the rose, true lily, and the other flowers of luxury. And that the deeply underlying reason of this is in the relation of weeds to corn, or of the adverse powers of nature to the beneficent ones, expressed for us readers of the Jewish scriptures, centrally in the verse, "thorns also, and thistles, shall it bring forth to thee; and thou shalt eat the herb of the field" (χορτος, grass or corn), and exquisitely symbolized throughout the fields of Europe by the presence of the purple 'corn-flag,' or gladiolus, and 'corn-rose' (Gerarde's name for Papaver Rhoeas), in the midst of carelessly tended

corn; and in the traditions of the art of Europe by the springing of the acanthus round the basket of the canephora, strictly the basket *for bread*, the idea of bread including all sacred things carried at the feasts of Demeter, Bacchus, and the Queen of the Air. And this springing of the thorny weeds round the basket of reed, distinctly taken up by the Byzantine Italians in the basket-work capital of the twelfth century, (which I have already illustrated at length in the 'Stones of Venice,') becomes the germ of all capitals whatsoever, in the great schools of Gothic, to the end of Gothic time, and also of all the capitals of the pure and noble Renaissance architecture of Angelico and Perugino, and all that was learned from them in the north, while the introduction of the rose, as a primal element of decoration, only takes place when the luxury of English decorated Gothic, the result of that licentious spirit in the lords which brought on the Wars of the Roses, indicates the approach of destruction to the feudal, artistic, and moral power of the northern nations.

For which reason, and many others, I must yet delay the following out of our main subject, till I have answered the other question, which brought me to pause in the middle of this chapter, namely, 'What is a weed?'

CHAPTER VI.

THE PARABLE OF JOASH.

1. SOME ten or twelve years ago, I bought—three times twelve are thirty-six—of a delightful little book by Mrs. Gatty, called 'Aunt Judy's Tales'—whereof to make presents to my little lady friends. I had, at that happy time, perhaps from four-and-twenty to six-and-thirty—I forget exactly how many—very particular little lady friends; and greatly wished Aunt Judy to be the thirty-seventh,—the kindest, wittiest, prettiest girl one had ever read of, at least in so entirely proper and orthodox literature.

2. Not but that it is a suspicious sign of infirmity of faith in our modern moralists to make their exemplary young people always pretty; and dress them always in the height of the fashion. One may read Miss Edgeworth's 'Harry and Lucy,' 'Frank and Mary,' 'Fashionable Tales,' or 'Parents' Assistant,' through, from end to end, with extremest care; and never find out whether Lucy was tall or short, nor whether Mary was dark or fair, nor how

Miss Annaly was dressed, nor—which was my own chief point of interest—what was the colour of Rosamond's eyes. Whereas Aunt Judy, in charming position after position, is shown to have expressed all her pure evangelical principles with the prettiest of lips; and to have had her gown, though puritanically plain, made by one of the best modistes in London.

3. Nevertheless, the book is wholesome and useful; and the nicest story in it, as far as I recollect, is an inquiry into the subject which is our present business, 'What is a weed?'—in which, by many pleasant devices, Aunt Judy leads her little brothers and sisters to discern that a weed is 'a plant in the wrong place.'

'Vegetable' in the wrong place, by the way, I think Aunt Judy says, being a precisely scientific little aunt. But I can't keep it out of my own less scientific head that 'vegetable' means only something going to be boiled. I like 'plant' better for general sense, besides that it's shorter.

Whatever we call them, Aunt Judy is perfectly right about them as far as she has gone; but, as happens often even to the best of evangelical instructresses, she has stopped just short of the gist of the whole matter. It is entirely true that a weed is a plant that has got into a wrong place; but it never seems to have occurred to Aunt Judy that some plants never *do!*

Who ever saw a wood anemone or a heath blossom in the wrong place? Who ever saw nettle or hemlock in a right one? And yet, the difference between flower and weed, (I use, for convenience' sake, these words in their familiar opposition,) certainly does not consist merely in the flowers being innocent, and the weed stinging and venomous. We do not call the nightshade a weed in our hedges, nor the scarlet agaric in our woods. But we do the corncockle in our fields.

4. Had the thoughtful little tutress gone but one thought farther, and instead of "a vegetable in a wrong place," (which it may happen to the innocentest vegetable sometimes to be, without turning into a weed, therefore,) said, "A vegetable which has an innate disposition to *get* into the wrong place," she would have greatly furthered the matter for us; but then she perhaps would have felt herself to be uncharitably dividing with vegetables her own little evangelical property of original sin.

5. This, you will find, nevertheless, to be the very essence of weed character—in plants, as in men. If you glance through your botanical books, you will see often added after certain names—'a troublesome weed.' It is not its being venomous, or ugly, but

its being impertinent—thrusting itself where it has no business, and hinders other people's business—that makes a weed of it. The most accursed of all vegetables, the one that has destroyed for the present even the possibility of European civilization, is only called a weed in the slang of its votaries; * but in the finest and truest English we call so the plant which has come to us by chance from the same country, the type of mere senseless prolific activity, the American water-plant, choking our streams till the very fish that leap out of them cannot fall back, but die on the clogged surface; and indeed, for this unrestrainable, unconquerable insolence of uselessness, what name can be enough dishonourable?

6. I pass to vegetation of nobler rank.

You remember, I was obliged in the last chapter to leave my poppy, for the present, without an English specific name, because I don't like Gerarde's 'Corn-rose,' and can't yet think of another. Nevertheless, I would have used Gerarde's name, if the corn-rose were as much a rose as the corn-flag is a flag. But it isn't. The rose and lily have quite different relations to the corn. The lily is grass in loveliness,

* And I have too harshly called our English vines, 'wicked weeds of Kent,' in Fors Clavigera, xxvii. 11. Much may be said for Ale, when we brew it for our people honestly.

as the corn is grass in use; and both grow together in peace—gladiolus in the wheat, and narcissus in the pasture. But the rose is of another and higher order than the corn, and you never saw a cornfield overrun with sweetbriar or apple-blossom.

They have no mind, they, to get into the wrong place.

What is it, then, this temper in some plants—malicious as it seems—intrusive, at all events, or erring,—which brings them out of their places—thrusts them where they thwart us and offend?

7. Primarily, it is mere hardihood and coarseness of make. A plant that can live anywhere, will often live where it is not wanted. But the delicate and tender ones keep at home. You have no trouble in 'keeping down' the spring gentian. It rejoices in its own Alpine home, and makes the earth as like heaven as it can, but yields as softly as the air, if you want it to give place. Here in England, it will only grow on the loneliest moors, above the high force of Tees; its Latin name, for *us* (I may as well tell you at once) is to be 'Lucia verna;' and its English one, Lucy of Teesdale.

8. But a plant may be hardy, and coarse of make, and able to live anywhere, and yet be no weed. The coltsfoot, so far as I know, is the first of large-leaved plants to grow afresh on ground that has

been disturbed: fall of Alpine débris, run of railroad embankment, waste of drifted slime by flood, it seeks to heal and redeem; but it does not offend us in our gardens, nor impoverish us in our fields.

Nevertheless, mere coarseness of structure, indiscriminate hardihood, is at least a point of some unworthiness in a plant. That it should have no choice of home, no love of native land, is ungentle; much more if such discrimination as it has, be immodest, and incline it, seemingly, to open and much-traversed places, where it may be continually seen of strangers. The tormentilla gleams in showers along the mountain turf; her delicate crosslets are separate, though constellate, as the rubied daisy. But the king-cup—(blessing be upon it always no less)—crowds itself sometimes into too burnished flame of inevitable gold. I don't know if there was anything in the darkness of this last spring to make it brighter in resistance; but I never saw any spaces of full warm yellow, in natural colour, so intense as the meadows between Reading and the Thames; nor did I know perfectly what purple and gold meant, till I saw a field of park land embroidered a foot deep with king-cup and clover—while I was correcting my last notes on the spring colours of the Royal Academy—at Aylesbury.

9. And there are two other questions of extreme

subtlety connected with this main one. What shall we say of the plants whose entire destiny is parasitic—which are not only sometimes, and *im*pertinently, but always, and pertinently, out of place; not only out of the right place, but out of any place of their own? When is mistletoe, for instance, in the right place, young ladies, think you? On an apple tree, or on a ceiling? When is ivy in the right place?—when wallflower? The ivy has been torn down from the towers of Kenilworth; the weeds from the arches of the Coliseum, and from the steps of the Aracceli,—irreverently, vilely, and in vain; but how are we to separate the creatures whose office it is to abate the grief of ruin by their gentleness,

> "wafting wallflower scents
> From out the crumbling ruins of fallen pride,
> And chambers of transgression, now forlorn,"

from those which truly resist the toil of men, and conspire against their fame; which are cunning to consume, and prolific to encumber; and of whose perverse and unwelcome sowing we know, and can say assuredly, "An enemy hath done this."

10. Again. The character of strength which gives prevalence over others to any common plant, is more or less consistently dependent on woody fibre

in the leaves; giving them strong ribs and great expanding extent; or spinous edges, and wrinkled or gathered extent.

Get clearly into your mind the nature of these two conditions. When a leaf is to be spread wide, like the Burdock, it is supported by a framework of extending ribs like a Gothic roof. The supporting function of these is geometrical; every one is constructed like the girders of a bridge, or beams of a floor, with all manner of science in the distribution of their substance in the section, for narrow and deep strength; and the shafts are mostly hollow. But when the extending space of a leaf is to be enriched with fulness of folds, and become beautiful in wrinkles, this may be done either by pure undulation as of a liquid current along the leaf edge, or by sharp 'drawing'—or 'gathering' I believe ladies would call it—and stitching of the edges together. And this stitching together, if to be done very strongly, is done round a bit of stick, as a sail is reefed round a mast; and this bit of stick needs to be compactly, not geometrically strong; its function is essentially that of starch,—not to hold the leaf up off the ground against gravity; but to stick the edges out, stiffly, in a crimped frill. And in beautiful work of this kind, which we are meant to study, the stays of the leaf—or stay-bones—are finished off very sharply and exquisitely at the points; and indeed so much so, that they prick our fingers when we touch them; for they are not at all meant to be touched, but admired.

11. To be admired,—with qualification, indeed, always, but with extreme respect for their endurance and orderliness. Among flowers that pass away, and leaves that shake as with ague, or shrink like bad cloth,—these, in their sturdy growth and enduring life, we are bound to honour; and, under the green holly, remember how much softer friendship was failing, and how much of other loving, folly. And yet,—you are not to confuse the thistle with the cedar that is in Lebanon; nor to forget—if the spinous nature of it become too cruel to provoke and offend—the parable of Joash to Amaziah, and its fulfilment: "There passed by a wild beast that was in Lebanon, and trode down the thistle."

12. Then, lastly, if this rudeness and insensitiveness of nature be gifted with no redeeming beauty; if the boss of the thistle lose its purple, and the star of the Lion's tooth, its light; and, much more, if service be perverted as beauty is lost, and the honied tube, and medicinal leaf, change into mere swollen emptiness, and salt brown membrane, swayed in nerveless

Drawn by J. Ruskin. Engraved by G. Allen

III

Acanthoid Leaves.

NORTHERN ATTIC TYPE

languor by the idle sea,—at last the separation between the two natures is as great as between the fruitful earth and fruitless ocean; and between the living hands that tend the Garden of Herbs where Love is, and those unclasped, that toss with tangle and with shells.

* * * * *

13. I had a long bit in my head, that I wanted to write, about St. George of the Seaweed, but I've no time to do it; and those few words of Tennyson's are enough, if one thinks of them: only I see, in correcting press, that I've partly misapplied the idea of 'gathering' in the leaf edge. It would be more accurate to say it was gathered at the central rib; but there is nothing in needlework that will represent the actual excess by lateral growth at the edge, giving three or four inches of edge for one of centre. But the stiffening of the fold by the thorn which holds it out is very like the action of a ship's spars on its sails; and absolutely in many cases like that of the spines in a fish's fin, passing into the various conditions of serpentine and dracontic crest, connected with all the terrors and adversities of nature; not to be dealt with in a chapter on weeds.

14. Here is a sketch of a crested leaf of less adverse temper, which may as well be given, together with Plate III., in this number, these two

engravings being meant for examples of two different methods of drawing, both useful according to character of subject. Plate III. is sketched first with a finely-pointed pen, and common ink, on white paper: then washed rapidly with colour, and retouched with the pen to give sharpness and completion. This method is used because the thistle leaves are full of complex and sharp sinuosities, and set with intensely sharp spines passing into hairs, which require many kinds of execution with the fine point to imitate at all. In the drawing there was more look of the bloom or woolliness on the stems, but it was useless to try for this in the mezzotint, and I desired Mr. Allen to leave his work at the stage where it expressed as much form as I wanted. The leaves are of the common marsh thistle, of which more anon; and the two long lateral ones are only two different views of the same leaf, while the central figure is a young leaf just opening. It beat me, in its delicate bossing, and I had to leave it, discontentedly enough.

Plate IV. is much better work, being of an easier subject, adequately enough rendered by perfectly simple means. Here I had only a succulent and membranous surface to represent, with definite outlines, and merely undulating folds; and this is sufficiently

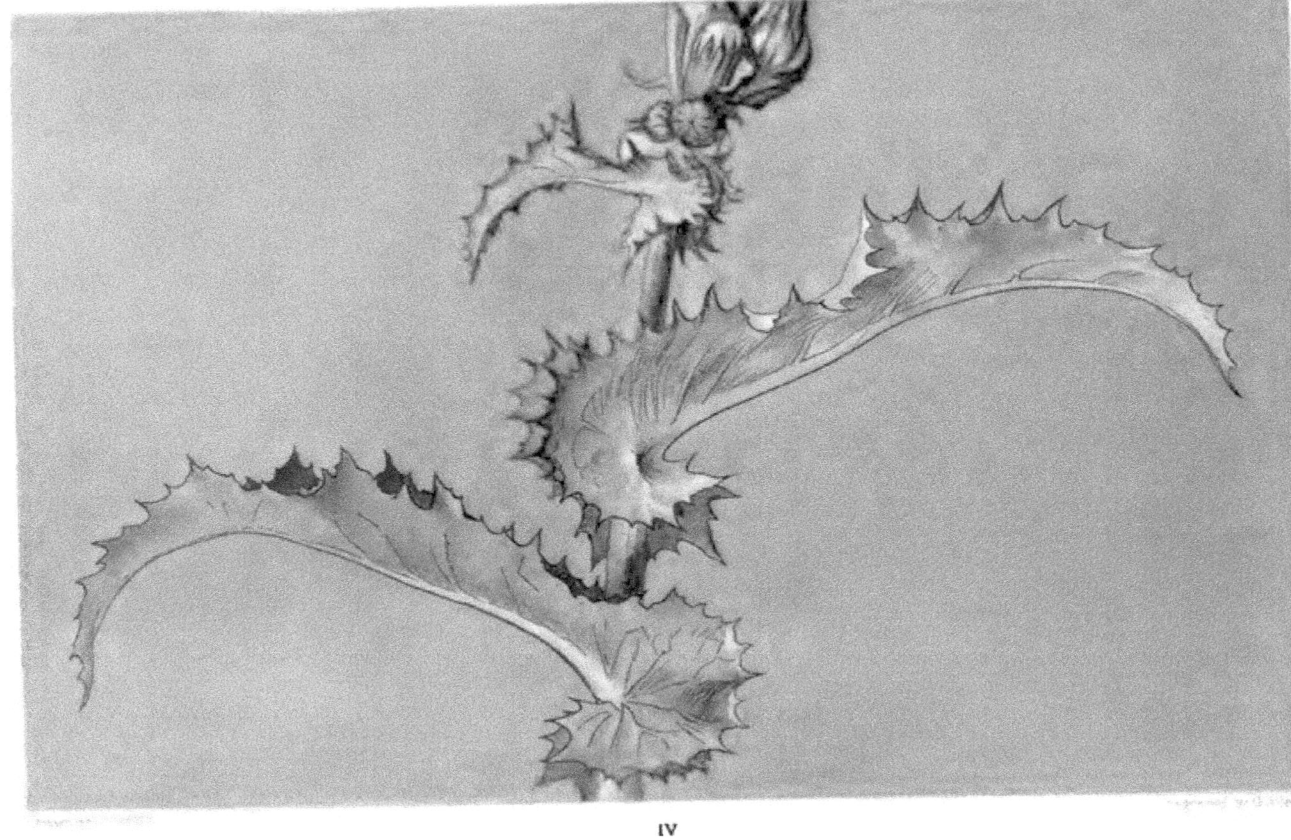

IV

done by a careful and firm pen outline on grey paper, with a slight wash of colour afterwards, reinforced in the darks; then marking the lights with white. This method is classic and authoritative, being used by many of the greatest masters, (by Holbein continually;) and it is much the best which the general student can adopt for expression of the action and muscular power of plants.

The goodness or badness of such work depends absolutely on the truth of the single line. You will find a thousand botanical drawings which will give you a delicate and deceptive resemblance of the leaf, for one that will give you the right convexity in its backbone, the right perspective of its peaks when they foreshorten, or the right relation of depth in the shading of its dimples. On which, in leaves as in faces, no little expression of temper depends.

Meantime we have yet to consider somewhat more touching that temper itself, in next chapter.

CHAPTER VII.

THE PARABLE OF JOTHAM.

1. I DO not know if my readers were checked, as I wished them to be, at least for a moment, in the close of the last chapter, by my talking of thistles and dandelions changing into seaweed, by gradation of which, doubtless, Mr. Darwin can furnish us with specious and sufficient instances. But the two groups will not be contemplated in our Oxford system as in any parental relations whatsoever.

We shall, however, find some very notable relations existing between the two groups of the wild flowers of dry land, which represent, in the widest extent, and the distinctest opposition, the two characters of material serviceableness and unserviceableness; the groups which in our English classification will be easily remembered as those of the Thyme, and the Daisy.

The one, scented as with incense—medicinal—and in all gentle and humble ways, useful. The other, scentless—helpless for ministry to the body;

infinitely dear as the bringer of light, ruby, white and gold ; the three colours of the Day, with no hue of shade in it. Therefore I take it on the coins of St. George for the symbol of the splendour or light of heaven, which is dearest where humblest.

2. Now these great two orders—of which the types are the thyme and the daisy—you are to remember generally as the 'Herbs' and the 'Sunflowers.' You are not to call them Lipped flowers, nor Composed flowers ; because the first is a vulgar term ; for when you once come to be able to draw a lip, or, in noble duty, to kiss one, you will know that no other flower in earth is like that : and the second is an indefinite term ; for a foxglove is as much a 'composed' flower as a daisy ; but it is composed in the shape of a spire, instead of the shape of the sun. And again a thistle, which common botany calls a composed flower, as well as a daisy, is composed in quite another shape, being, on the whole, bossy instead of flat ; and of another temper, or composition of mind, also, being connected in that respect with butterburs, and a vast company of rough, knotty, half-black or brown, and generally unluminous—flowers I can scarcely call them—and weeds I will not,—creatures, at all events, in nowise to be gathered under the general name 'Composed,' with the stars that crown Chaucer's

Alcestis, when she returns to the day from the dead.

But the wilder and stronger blossoms of the Hawk's-eye—again you see I refuse for them the word weed;—and the waste-loving Chicory, which the Venetians call 'Sponsa solis,' are all to be held in one class with the Sunflowers; but dedicate,—the daisy to Alcestis alone; others to Clytia, or the Physician Apollo himself: but I can't follow their mythology yet awhile.

3. Now in these two families you have typically Use opposed to Beauty in *wildness;* it is their wildness which is their virtue;—that the thyme is sweet where it is unthought of, and the daisies red, where the foot despises them: while, in other orders, wildness is their crime,—"Wherefore, when I looked that it should bring forth grapes, brought it forth wild grapes?" But in all of them you must distinguish between the pure wildness of flowers and their distress. It may not be our duty to tame them; but it must be, to relieve.

4. It chanced, as I was arranging the course of these two chapters, that I had examples given me of distressed and happy wildness, in immediate contrast. The first, I grieve to say, was in a bit of my own brushwood, left uncared-for evidently

many a year before it became mine. I had to cut my way into it through a mass of thorny ruin; black, bird's-nest like, entanglement of brittle spray round twisted stems of ill-grown birches strangling each other, and changing half into roots among the rock clefts; knotted stumps of never-blossoming blackthorn, and choked stragglings of holly, all laced and twisted and tethered round with an untouchable, almost unhewable, thatch, a foot thick, of dead bramble and rose, laid over rotten ground through which the water soaked ceaselessly, undermining it into merely unctuous clods and clots, knitted together by mossy sponge. It was all Nature's free doing! she had had her way with it to the uttermost; and clearly needed human help and interference in her business; and yet there was not one plant in the whole ruinous and deathful riot of the place, whose nature was not in itself wholesome and lovely; but all lost for want of discipline.

5. The other piece of wild growth was among the fallen blocks of limestone under Malham Cove. Sheltered by the cliff above from stress of wind, the ash and hazel wood spring there in a fair and perfect freedom, without a diseased bough, or an unwholesome shade. I do not know why mine

is all encumbered with overgrowth, and this so lovely that scarce a branch could be gathered but with injury ;—while underneath, the oxalis, and the two smallest geraniums (Lucidum and Herb-Robert) and the mossy saxifrage, and the cross-leaved bed-straw, and the white pansy, wrought themselves into wreaths among the fallen crags, in which every leaf rejoiced, and was at rest.

6. Now between these two states of equally natural growth, the point of difference that forced itself on me (and practically enough, in the work I had in my own wood), was not so much the withering and waste of the one, and the life of the other, as the thorniness and cruelty of the one, and the softness of the other. In Malham Cove, the stones of the brook were softer with moss than any silken pillow—the crowded oxalis leaves yielded to the pressure of the hand, and were not felt—the cloven leaves of the Herb-Robert and orbed clusters of its companion overflowed every rent in the rude crags with living balm ; there was scarcely a place left by the tenderness of the happy things, where one might not lay down one's forehead on their warm softness, and sleep. But in the waste and distressed ground, the distress had changed itself to cruelty. The leaves had all perished, and the bending saplings, and the wood of trust ;—but the thorns

were there, immortal, and the gnarled and sapless roots, and the dusty treacheries of decay.

7. Of which things you will find it good to consider also otherwise than botanically. For all these lower organisms suffer and perish, or are gladdened and flourish, under conditions which are in utter precision symbolical, and in utter fidelity representative, of the conditions which induce adversity and prosperity in the kingdoms of men: and the Eternal Demeter,—Mother, and Judge,—brings forth, as the herb yielding seed, so also the thorn and the thistle, not to herself, but *to thee*.

8. You have read the words of the great Law often enough;—have you ever thought enough of them to know the difference between these two appointed means of Distress? The first, the Thorn, is the type of distress *caused by crime*, changing the soft and breathing leaf into inflexible and wounding stubbornness. The second is the distress appointed to be the means and herald of good,—Thou shalt see the stubborn thistle bursting into glossy purple, which outreddens all voluptuous garden roses.

9. It is strange that, after much hunting, I cannot find authentic note of the day when Scotland took the thistle for her emblem; and I have no space (in this chapter at least) for tradition; but, with

whatever lightness of construing we may receive the symbol, it is actually the truest that could have been found, for some conditions of the Scottish mind. There is no flower which the Proserpina of our Northern Sicily cherishes more dearly: and scarcely any of us recognize enough the beautiful power of its close-set stars, and rooted radiance of ground leaves; yet the stubbornness and ungraceful rectitude of its stem, and the besetting of its wholesome substance with that fringe of offence, and the forwardness of it, and dominance,—I fear to lacess some of my dearest friends if I went on:—let them rather, with Bailie Jarvie's true conscience,* take their Scott from the inner shelf in their heart's library which all true Scotsmen give him, and trace, with the swift reading of memory, the characters of Fergus M'Ivor, Hector M'Intyre, Mause Headrigg, Alison Wilson, Richie Moniplies, and Andrew Fairservice; and then say, if the faults of all these, drawn as they are with

* Has my reader ever thought,—I never did till this moment,—how it perfects the exquisite character which Scott himself loved, as he invented, till he changed the form of the novel, that his habitual interjection should be this word?—not but that the oath, by conscience, was happily still remaining then in Scotland, taking the place of the mediæval 'by St. Andrew,' we in England, long before the Scot, having lost all sense of the Puritanical appeal to private conscience, as of the Catholic oath, 'by St. George;' and our uncanonized 'by George' in sonorous rudeness, ratifying, not now our common conscience, but our individual opinion.

a precision of touch like a Corinthian sculptor's of the acanthus leaf, can be found in anything like the same strength in other races, or if so stubbornly folded and starched moni-plies of irritating kindliness, selfish friendliness, lowly conceit, and intolerable fidelity, are native to any other spot of the wild earth of the habitable globe.

10. Will you note also—for this is of extreme interest—that these essential faults are all mean faults;—what we may call ground-growing faults; conditions of semi-education, of hardly-treated home-life, or of coarsely-minded and wandering prosperity? How literally may we go back from the living soul symbolized, to the strangely accurate earthly symbol, in the prickly weed. For if, with its bravery of endurance, and carelessness in choice of home, we find also definite faculty and habit of migration, volant mechanism for choiceless journey, not divinely directed in pilgrimage to known shrines; but carried at the wind's will by a spirit which listeth *not*,—it will go hard but that the plant shall become, if not dreaded, at least despised; and, in its wandering and reckless splendour, disgrace the garden of the sluggard, and possess the inheritance of the prodigal: until even its own nature seems contrary to good, and the invocation of the just man be made to it as the executor of Judgment,

"Let thistles grow instead of wheat, and cockle instead of barley."

11. Yet to be despised—either for men or flowers—may be no ill-fortune; the real ill-fortune is only to be despicable. These faults of human character, wherever found, observe, belong to it as ill-trained—incomplete; confirm themselves only in the vulgar. There is no base pertinacity, no overweening conceit, in the Black Douglas, or Claverhouse, or Montrose; in these we find the pure Scottish temper, of heroic endurance and royal pride; but, when, in the pay, and not deceived, but purchased, idolatry of Mammon, the Scottish persistence and pride become knit and vested in the spleuchan, and your stiff Covenanter makes his covenant with Death, and your Old Mortality deciphers only the senseless legends of the eternal gravestone,—you get your weed, earth-grown, in bitter verity, and earth-devastating, in bitter strength.

12. I have told you elsewhere, we are always first to study national character in the highest and purest examples. But if our knowledge is to be complete, we have to study also the special diseases of national character. And in exact opposition to the most solemn virtue of Scotland, the domestic truth and tenderness breathed in all Scottish song, you have this special disease and mortal cancer, this woody-fibriness, literally, of temper and thought:

the consummation of which into pure lignite, or rather black Devil's charcoal—the sap of the birks of Aberfeldy become cinder, and the blessed juices of them, deadly gas,—you may know in its pure blackness best in the work of the greatest of these ground-growing Scotchmen, Adam Smith.

13. No man of like capacity, I believe, born of any other nation, could have deliberately, and with no momentary shadow of suspicion or question, formalized the spinous and monstrous fallacy that human commerce and policy are *naturally* founded on the desire of every man to possess his neighbour's goods.

This is the 'release unto us Barabbas,' with a witness; and the deliberate systematization of that cry, and choice, for perpetual repetition and fulfilment in Christian statesmanship, has been, with the strange precision of natural symbolism and retribution, signed, (as of old, by strewing of ashes on Kidron,) by strewing of ashes on the brooks of Scotland; waters once of life, health, music, and divine tradition; but to whose festering scum you may now set fire with a candle; and of which, round the once excelling palace of Scotland, modern sanitary science is now helplessly contending with the poisonous exhalation.

14. I gave this chapter its heading, because I had it

in my mind to work out the meaning of the fable in the ninth chapter of Judges, from what I had seen on that thorny ground of mine, where the bramble was king over all the trees of the wood. But the thoughts are gone from me now; and as I re-read the chapter of Judges,—now, except in my memory, unread, as it chances, for many a year,—the sadness of that story of Gideon fastens on me, and silences me. *This* the end of his angel visions, and dream-led victories, the slaughter of all his sons but this youngest,*—and he never again heard of in Israel!

You Scottish children of the Rock, taught through all your once pastoral and noble lives by many a sweet miracle of dew on fleece and ground,—once servants of mighty kings, and keepers of sacred covenant; have you indeed dealt truly with your warrior kings, and prophet saints, or are these ruins of their homes, and shrines, dark with the fire that fell from the curse of Jerubbael?

* 'Jotham,' 'Sum perfectio eorum,' or 'Consummatio eorum.' (Interpretation of name in Vulgate Index.)

CHAPTER VIII.

THE STEM.

1. AS I read over again, with a fresh mind, the last chapter, I am struck by the opposition of states which seem best to fit a weed for a weed's work,—stubbornness, namely, and flaccidity. On the one hand, a sternness and a coarseness of structure which changes its stem into a stake, and its leaf into a spine; on the other, an utter flaccidity and ventosity of structure, which changes its stem into a riband, and its leaf into a bubble. And before we go farther—for we are not yet at the end of our study of these obnoxious things—we had better complete an examination of the parts of a plant in general, by ascertaining what a Stem proper is; and what makes it stiffer, or hollower, than we like it;—how, to wit, the gracious and generous strength of ash differs from the spinous obstinacy of blackthorn,—and how the geometric and enduring hollowness of a stalk of wheat differs from the soft fulness of that of a mushroom. To which end, I will take up a piece of study, not of black, but white, thorn, written last spring.

2. I suppose there is no question but that all nice people like hawthorn blossom.

I want, if I can, to find out to-day, 25th May, 1875, what it is we like it so much for: holding these two branches of it in my hand,—one full out, the other in youth. This full one is a mere mass of symmetrically balanced—snow, one was going vaguely to write, in the first impulse. But it is nothing of the sort. White,—yes, in a high degree; and pure, totally; but not at all dazzling in the white, nor pure in an insultingly rivalless manner, as snow would be; yet pure somehow, certainly; and white, absolutely, in spite of what might be thought failure,—imperfection—nay, even distress and loss in it. For every little rose of it has a green darkness in the centre—not even a pretty green, but a faded, yellowish, glutinous, unaccomplished green; and round that, all over the surface of the blossom, whose shell-like petals are themselves deep sunk, with grey shadows in the hollows of them—all above this already subdued brightness, are strewn the dark points of the dead stamens—manifest more and more, the longer one looks, as a kind of grey sand, sprinkled without sparing over what looked at first unspotted light. And in all the ways of it the lovely thing is more like the spring frock of some prudent little maid of fourteen, than a flower;

—frock with some little spotty pattern on it to keep it from showing an unintended and inadvertent spot—if Fate should ever inflict such a thing! Undeveloped, thinks Mr. Darwin,—the poor short-coming, ill-blanched thorn blossom—going to be a Rose, some day soon; and, what next?—who knows?—perhaps a Pæony!

3. Then this next branch, in dawn and delight of youth, set with opening clusters of yet numerable blossom, four, and five, and seven, edged, and islanded, and ended, by the sharp leaves of freshest green, deepened under the flowers, and studded round with bosses, better than pearl beads of St. Agnes' rosary,—folded, over and over, with the edges of their little leaves pouting, as the very softest waves do on flat sand where one meets another; then opening just enough to show the violet colour within—which yet isn't violet colour, nor even "meno che le rose," but a different colour from every other lilac that one ever saw;—faint and faded even before it sees light, as the filmy cup opens over the depth of it, then broken into purple motes of tired bloom, fading into darkness, as the cup extends into the perfect rose.

This, with all its sweet change that one would so fain stay, and soft effulgence of bud into softly falling flower, one has watched—how often; but always with the feeling that the blossoms are thrown over the green depth like white clouds,—never with any idea of so much as asking what holds the cloud there. Have each of the innumerable blossoms a separate stalk; and, if so, how is it that one never thinks of the stalk, as one does with currants?

4. Turn the side of the branch to you;—Nature never meant you to see it so; but now it is all stalk below and stamens above,—the petals nothing, the stalks all tiny trees, always dividing their branches mainly into three—one in the centre short, and the two lateral, long, with an intermediate extremely long one, if needed, to fill a gap, so contriving that the flowers shall all be nearly at the same level, or at least surface of ball, like a guelder rose. But the cunning with which the tree conceals its structure till the blossom is fallen, and then—for a little while, we had best look no more at it, for it is all like grape-stalks with no grapes.

These, whether carrying hawthorn blossom and haw, or grape blossom and grape, or peach blossom and peach, you will simply call the 'stalk,' whether of flower or fruit. A 'stalk' is essentially round, like a pillar; and has, for the most part, the power of first developing, and then shaking off, flower and fruit from its extremities. You can pull the peach from its stalk, the cherry, the grape. Always at some time of its existence, the flower-stalk lets fall something of what is sustained, petal or seed.

In late Latin it is called 'petiolus,' the little foot; because the expanding piece that holds the grape, or olive, is a little like an animal's foot. Modern botanists have misapplied the word to the *leaf*-stalk, which has no resemblance to a foot at all. We must keep the word to its proper meaning, and, when we want to write Latin, call it 'petiolus;' when we want to write English, call it 'stalk,' meaning always fruit or flower stalk.

I cannot find when the word 'stalk' first appears in English:—its derivation will be given presently.

5. Gather next a hawthorn leaf. That also has a stalk; but you can't shake the leaf off it. It, and the leaf, are essentially one; for the sustaining fibre runs up into every ripple or jag of the leaf's edge: and its section is different from that of the flower-stalk; it is no more round, but has an upper and under surface, quite different from each other. It will be better, however, to take a larger leaf to examine this structure in. Cabbage, cauliflower, or rhubarb, would any of them be good, but don't grow wild in the luxuriance I want. So, if you please, we will take a leaf of burdock, (Arctium Lappa,) the

principal business of that plant being clearly to grow leaves wherewith to adorn fore-grounds.*

6. The outline of it in Sowerby is not an intelligent one, and I have not time to draw it but in the rudest way myself; Fig. 13, *a*; with perspectives of the elementary form below, *b*, *c*, and *d*. By help of which, if you will construct a burdock leaf in paper, my rude outline (*a*) may tell the rest of what I want you to see.

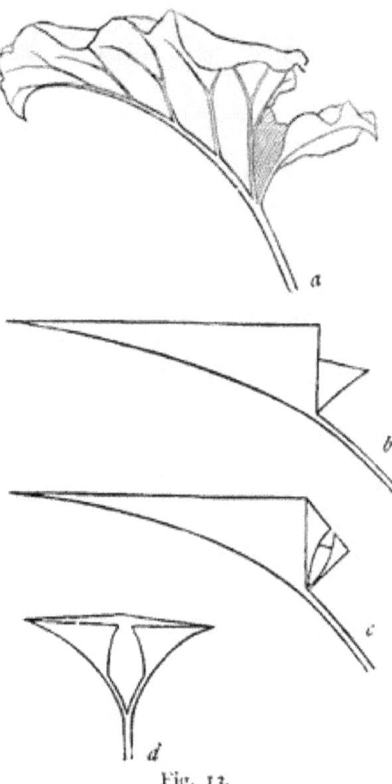

Fig. 13.

Take a sheet of stout note paper, Fig. 14, A, double it sharply down the centre, by the dotted line, then

* If you will look at the engraving, in the England and Wales series, of Turner's Oakhampton, you will see its use.

give it the two cuts at *a* and *b*, and double those pieces sharply back, as at B; then, opening them again, cut the whole into the form C; and then, pulling up the corners *c d*, stitch them together with a loose thread so that the points

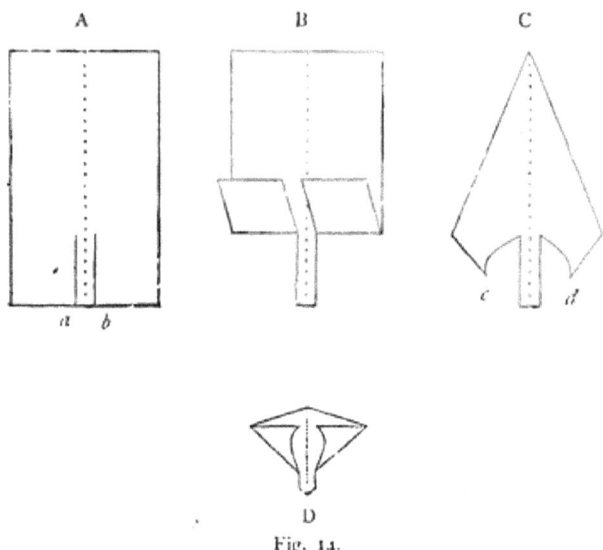

Fig. 14.

c and *d* shall be within half an inch of each other; and you will have a kind of triangular scoop, or shovel, with a stem, by which you can sufficiently hold it, D.

7. And from this easily constructed and tenable model, you may learn at once these following main facts about all leaves.

[I.] That they are not flat, but, however slightly, always hollowed into craters, or raised into hills, in one or another direction; so that any drawable outline of them does not in the least represent the real extent of their surfaces; and until you know how to draw a cup, or a mountain, rightly, you have no chance of drawing a leaf. My simple artist readers of long ago, when I told them to draw leaves, thought they could do them by the bough-full, whenever they liked. Alas, except by old William Hunt, and Burne Jones, I've not seen a leaf painted, since those burdocks of Turner's; far less sculptured—though one would think at first that was easier! Of which we shall have talk elsewhere; here I must go on to note fact number two, concerning leaves.

8. [II.] The strength of their supporting stem consists not merely in the gathering together of all the fibres, but in gathering them essentially into the profile of the letter V, which you will see your doubled paper stem has; and of which you can feel the strength and use, in your hand, as you hold it. Gather a common plantain leaf, and look at the way it puts its round ribs together at the base, and you will understand the matter at once. The arrangement is modified and

disguised in every possible way, according to the leaf's need: in the aspen, the leaf-stalk becomes an absolute vertical plank; and in the large trees is often almost rounded into the likeness of a fruit-stalk;—but, in all,* the essential structure is this doubled one; and in all, it opens at the place where the leaf joins the main stem, into a kind of cup, which holds next year's bud in the hollow of it.

9. Now there would be no inconvenience in your simply getting into the habit of calling the round petiol of the fruit the 'stalk,' and the contracted channel of the leaf, 'leaf-stalk.' But this way of naming them would not enforce, nor fasten in your mind, the difference between the two, so well as if you have an entirely different name for the leaf-stalk. Which is the more desirable, because the limiting character of the leaf, botanically, is—(I only learned this from my botanical friend the other day, just in the very moment I wanted it,)—that it holds the bud of the new stem in its own hollow, but cannot itself grow in the hollow of anything else;—or, in botanical language, leaves are never axillary,—don't grow in armpits, but are themselves armpits; hollows,

* General assertions of this kind must always be accepted under indulgence,—exceptions being made afterwards.

that is to say, where they spring from the main stem.

10. Now there is already a received and useful botanical word, 'cyme' (which we shall want in a little while,) derived from the Greek κῦμα, a swelling or rising wave, and used to express a swelling cluster of foamy blossom. Connected with that word, but in a sort the reverse of it, you have the Greek 'κύμβη,' the *hollow* of a cup, or bowl; whence κύμβαλον, cymbal,—that is to say, a musical instrument owing its tone to its *hollowness*. These words become in Latin, cymba, and cymbalum; and I think you will find it entirely convenient and advantageous to call the leaf-stalk distinctively the 'cymba,' retaining the mingled idea of cup and boat, with respect at least to the part of it that holds the bud; and understanding that it gathers itself into a V-shaped, or even narrowly vertical, section, as a boat narrows to its bow, for strength to sustain the leaf.

With this word you may learn the Virgilian line, that shows the final use of iron—or iron-darkened—ships:

"Et ferrugineâ subvectat corpora cymbâ."

The "subvectat corpora" will serve to remind you

of the office of the leafy cymba in carrying the bud; and make you thankful that the said leafy vase is not of iron; and is a ship of Life instead of Death.

11. Already, not once, nor twice, I have had to use the word 'stem,' of the main round branch from which both stalk and cymba spring. This word you had better keep for all growing, or advancing, shoots of trees, whether from the ground, or from central trunks and branches. I regret that the words multiply on us; but each that I permit myself to use has its own proper thought or idea to express, as you will presently perceive; so that true knowledge multiplies with true words.

12. The 'stem,' you are to say, then, when you mean the *advancing* shoot,—which lengthens annually, while a stalk ends every year in a blossom, and a cymba in a leaf. A stem is essentially round,* square, or regularly polygonal; though, as a cymba may become exceptionally round, a stem may become exceptionally flat, or even mimic the shape of a leaf. Indeed I should have liked to write "a stem is essentially round, and constructively, on occasion, square,"—but it would have been too grand. The fact is, however, that a stem is really a roundly minded thing, throwing off its

* I use 'round' rather than 'cylindrical,' for simplicity's sake.

branches in circles as a trundled mop throws off drops, though it can always order the branches to fly off in what order it likes,—two at a time, opposite to each other; or three, or five, in a spiral coil; or one here and one there, on this side and that; but it is always twisting, in its own inner mind and force; hence it is especially proper to use the word 'stem' of it—*στέμμα*, a twined wreath; properly, twined round a staff, or sceptre: therefore, learn at once by heart these lines in the opening Iliad:

> "*Στέμματ' ἔχων ἐν χερσὶν ἑκηβόλου Ἀπόλλωνος,*
> *Χρυσέῳ ἀνὰ σκήπτρῳ·*"

And recollect that a sceptre is properly a staff to lean upon; and that as a crown or diadem is first a binding thing, a 'sceptre' is first a *supporting* thing, and it is in its nobleness, itself made of the stem of a young tree. You may just as well learn also this:

> "*Ναὶ μὰ τόδε σκῆπτρον, τὸ μὲν οὔποτε φύλλα καὶ ὄζους*
> *Φύσει, ἐπειδὴ πρῶτα τομὴν ἐν ὄρεσσι λέλοιπεν,*
> *Οὐδ' ἀναθηλήσει· περὶ γάρ ῥά ἑ χαλκὸς ἔλεψε*
> *Φύλλα τε καὶ φλοιόν· νῦν αὖτε μιν υἷες Ἀχαιῶν*
> *Ἐν παλάμης φορέουσι δικασπόλοι, οἵ τε θέμιστας*
> *Πρὸς Διὸς εἰρύαται·*"

"Now, by this sacred sceptre hear me swear
Which never more shall leaves or blossoms bear,
Which, severed from the trunk, (as I from thee,)
On the bare mountains left its parent tree;
This sceptre, formed by tempered steel to prove
An ensign of the delegates of Jove,
From whom the power of laws and justice springs
(Tremendous oath, inviolate to Kings)."

13. The supporting power in the tree itself is, I doubt not, greatly increased by this spiral action; and the fine instinct of its being so, caused the twisted pillar to be used in the Lombardic Gothic, —at first, merely as a pleasant variety of form, but at last constructively and universally, by Giotto, and all the architects of his school. Not that the spiral form actually adds to the strength of a Lombardic pillar, by imitating contortions of wood, any more than the fluting of a Doric shaft adds to its strength by imitating the canaliculation of a reed; but the perfect action of the imagination, which had adopted the encircling acanthus for the capital, adopted the twining stemma for the shaft; the pure delight of the eye being the first condition in either case: and it is inconceivable how much of the pleasure taken both in ornament and in natural form is founded elementarily on groups

of spiral line. The study, in our fifth plate, of the involucre of the waste-thistle,* is as good an example as I can give of the more subtle and concealed conditions of this structure.

14. Returning to our present business of nomenclature, we find the Greek word, 'stemma,' adopted by the Latins, becoming the expression of a growing and hereditary race; and the branched tree, the natural type, among all nations, of multiplied families. Hence the entire fitness of the word for our present purposes; as signifying, "a spiral shoot extending itself by branches." But since, unless it is spiral, it is not a stem, and unless it has branches, it is not a stem, we shall still want another word for the sustaining 'sceptre' of a foxglove, or cowslip. Before determining that, however, we must see what need there may be of one familiar to our ears until lately, although now, I understand, falling into disuse.

* Carduus Arvensis. 'Creeping Thistle,' in Sowerby; why, I cannot conceive, for there is no more creeping in it than in a furze-bush. But it especially haunts foul and neglected ground; so I keep the Latin name, translating 'Waste-Thistle.' I could not show the variety of the curves of the involucre without enlarging; and if, on this much increased scale, I had tried to draw the flower, it would have taken Mr. Allen and me a good month's more work. And I had no more a month than a life, to spare: so the action only of the spreading flower is indicated, but the involucre drawn with precision.

Engraved by G. Allen

V.

Occult Spiral Action.

WASTE-THISTLE.

15. By our definition, a stem is a spirally bent, essentially living and growing, shoot of vegetation. But the branch of a tree, in which many such stems have their origin, is not, except in a very subtle and partial way, spiral; nor except in the shoots that spring from it, progressive forwards; it only receives increase of thickness at its sides. Much more, what used to be called the *trunk* of a tree, in which many branches are united, has ceased to be, except in mere tendency and temper, spiral; and has so far ceased from growing as to be often in a state of decay in its interior, while the external layers are still in serviceable strength.

16. If, however, a trunk were only to be defined as an arrested stem, or a cluster of arrested stems, we might perhaps refuse, in scientific use, the popular word. But such a definition does not touch the main idea. Branches usually begin to assert themselves at a height above the ground approximately fixed for each species of tree,—low in an oak, high in a stone pine; but, in both, marked as a point of *structural change in the direction of growing force*, like the spring of a vault from a pillar; and as the tree grows old, some of its branches getting torn away by winds or falling under the weight of their own fruit, or load of snow, or by natural decay, there remains

literally a 'truncated' mass of timber, still bearing irregular branches here and there, but inevitably suggestive of resemblance to a human body, after the loss of some of its limbs.

And to prepare trees for their practical service, what age and storm only do partially, the first rough process of human art does completely. The branches are lopped away, leaving literally the 'truncus' as the part of the tree out of which log and rafter can be cut. And in many trees, it would appear to be the chief end of their being to produce this part of their body on a grand scale, and of noble substance; so that, while in thinking of vegetable life without reference to its use to men or animals, we should rightly say that the essence of it was in leaf and flower—not in trunk or fruit; yet for the sake of animals, we find that some plants, like the vine, are apparently meant chiefly to produce fruit; others, like laurels, chiefly to produce leaves; others chiefly to produce flowers; and others to produce permanently serviceable and sculpturable wood; or, in some cases, merely picturesque and monumental masses of vegetable rock, "intertwisted fibres serpentine,"—of far nobler and more pathetic use in their places, and their enduring age, than ever they could be for material purpose in human habitation. For this central mass of the vegetable organism, then, the English word 'trunk' and French 'tronc' are always in accurate scholarship to be retained—meaning the part of a tree which remains when its branches are lopped away.

17. We have now got distinct ideas of four different kinds of stem, and simple names for them in Latin and English,—Petiolus, Cymba, Stemma, and Truncus; Stalk, Leaf-stalk, Stem, and Trunk; and these are all that we shall commonly need. There is, however, one more that will be sometimes necessary, though it is ugly and difficult to pronounce, and must be as little used as we can.

And here I must ask you to learn with me a little piece of Roman history. I say, to *learn* with me, because I don't know any Roman history except the two first books of Livy, and little bits here and there of the following six or seven. I only just know enough about it to be able to make out the bearings and meaning of any fact that I now learn. The greater number of modern historians know, (if honest enough even for that,) the facts, or something that may possibly be like the facts, but haven't the least notion of the meaning of them. So that, though I have to find out everything that I want in Smith's Dictionary, like any schoolboy, I can usually tell you the significance of what I so find, better than perhaps even Mr. Smith himself could.

18. In the 586th page of Mr. Smith's volume, you have it written that 'Calvus,' bald-head, was the name of a family of the Licinia gens; that the man of whom we hear earliest, as so named, was the first plebeian elected to military tribuneship in B.C. 400; and that the fourth of whom we hear, was surnamed 'Stolo,' because he was so particular in pruning away the Stolons (stolones), or useless young shoots, of his vines.

We must keep this word 'stolon,' therefore, for these young suckers springing from an old root. Its derivation is uncertain; but the main idea meant by it is one of uselessness—sprouting without occasion or fruit; and the words 'stolidus' and 'stolid' are really its derivatives, though we have lost their sense in English by partly confusing them with 'solid,' which they have nothing to do with. A 'stolid' person is essentially a 'useless sucker' of society; frequently very leafy and graceful, but with no good in him.

19. Nevertheless, I won't allow our vegetable 'stolons' to be despised. Some of quite the most beautiful forms of leafage belong to them;—even the foliage of the olive itself is never seen to the same perfection on the upper branches as in

the young ground-rods in which the dual groups of leaves crowd themselves in their haste into clusters of three.

But, for our point of Latin history, remember always that in 400 B.C., just a year before the death of Socrates at Athens, this family of Stolid persons manifested themselves at Rome, shooting up from plebeian roots into places where they had no business; and preparing the way for the degradation of the entire Roman race under the Empire; their success being owed, remember also, to the faults of the patricians, for one of the laws passed by Calvus Stolo was that the Sibylline books should be in custody of ten men, of whom five should be plebeian, "that no falsifications might be introduced in favour of the patricians."

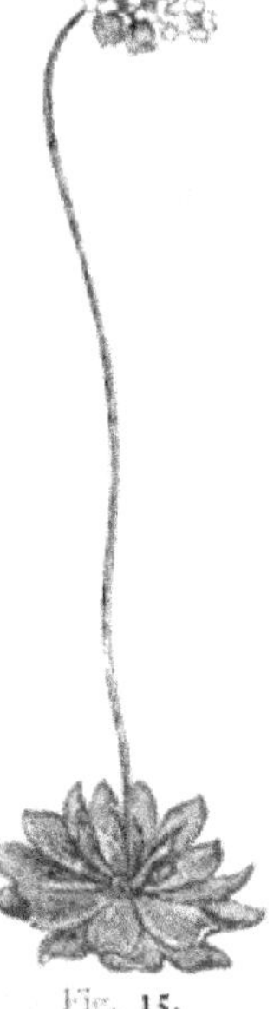
Fig. 15.

20. All this time, however, we have got no name for the prettiest of all stems,—that of annual flowers growing high from among their ground leaves, like lilies of the valley, and saxifrages, and the tall primulas—of which this pretty type, Fig. 15, was cut for me by Mr. Burgess years

ago; admirable in its light outline of the foamy globe of flowers, supported and balanced in the meadow breezes on that elastic rod of slenderest life.

What shall we call it? We had better rest from our study of terms a little, and do a piece of needful classifying, before we try to name it.

21. My younger readers will find it easy to learn, and convenient to remember, for a beginning of their science, the names of twelve great families of cinquefoiled flowers,* of which the first group of three, is for the most part golden, the second, blue, the third, purple, and the fourth, red.

And their names, by simple lips, can be pleasantly said, or sung, in this order, the two first only being a little difficult to get over.

1	2	3	4
Roof-foil,	Lucy,	Pea,	Pink,
Rock-foil,	Blue-bell,	Pansy,	Peach,
Primrose.	Bindweed.	Daisy.	Rose.

Which even in their Latin magniloquence will not be too terrible, namely,—

* The florets gathered in the daisy are cinquefoils, examined closely. No system founded on colour can be very general or unexceptionable: but the splendid purples of the pansy, and thistle, which will be made one of the lower composite groups under Margarita, may justify the general assertion of this order's being purple.

1	2	3	4
Stella,	Lucia,	Alata,	Clarissa,
Francesca,	Campanula,	Viola,	Persica,
Primula.	Convoluta.	Margarita.	Rosa.

22. I do not care much to assert or debate my reasons for the changes of nomenclature made in this list. The most gratuitous is that of 'Lucy' for 'Gentian,' because the King of Macedon, from whom the flower has been so long named, was by no means a person deserving of so consecrated memory. I conceive no excuse needed for rejecting Caryophyll, one of the crudest and absurdest words ever coined by unscholarly men of science; or Papilionaceæ, which is unendurably long for pease; and when we are now writing Latin, in a sentimental temper, and wish to say that we gathered a daisy, we shall not any more be compelled to write that we gathered a 'Bellidem perennem,' or, an 'Oculum Diei.'

I take the pure Latin form, Margarita, instead of Margareta, in memory of Margherita of Cortona,* as well as of the great saint: also the tiny scatterings and sparklings of the daisy on the turf may remind us of the old use of the word 'Margaritæ,' for

* See Miss Yonge's exhaustive account of the name, 'History of Christian Names,' vol. i., p. 265.

the minute particles of the Host sprinkled on the patina—"Has particulas μερίδας vocat Euchologium, μαργαρίτας Liturgia Chrysostomi."* My young German readers will, I hope, call the flower Gretschen,—unless they would uproot the daisies of the Rhine, lest French girls should also count their love-lots by the Marguerite. I must be so ungracious to my fair young readers, however, as to warn them that this trial of their lovers is a very favourable one, for, in nine blossoms out of ten, the leaves of the Marguerite are odd, so that, if they are only gracious enough to begin with the supposition that he loves them, they must needs end in the conviction of it.

23. I am concerned, however, for the present, only with my first or golden order, of which the Roof-foil, or house-leek, is called in present botany, Sedum, 'the squatter,' because of its way of fastening itself down on stones, or roof, as close as it can sit. But I think this an ungraceful notion of its behaviour; and as its blossoms are, of all flowers, the most sharply and distinctly star-shaped, I shall call it 'Stella' (providing otherwise, in due time, for the poor little chickweeds;) and

* (Du Cange.) The word 'Margarete' is given as heraldic English for pearl, by Lady Juliana Berners, in the book of St. Albans.

the common stonecrop will therefore be 'Stella domestica.'

The second tribe, (at present saxifraga,) growing for the most part wild on rocks, may, I trust, even in Protestant botany, be named Francesca, after St. Francis of Assisi; not only for its modesty, and love of mountain ground, and poverty of colour and leaf; but also because the chief element of its decoration, seen close, will be found in its spots, or stigmata.

In the nomenclature of the third order I make no change.

24. Now all this group of golden-blossoming plants agree in general character of having a rich cluster of radical leaves, from which they throw up a single stalk bearing clustered blossoms; for which stalk, when entirely leafless, I intend always to keep the term 'virgula,' the 'little rod'—not painfully caring about it, but being able thus to define it with precision, if required. And these are connected with the stems of branching shrubs through infinite varieties of structure, in which the first steps of transition are made by carrying the cluster of radical leaves up, and letting them expire gradually from the rising stem: the changes of form in the leaves as they rise higher from the ground being one of quite the most interesting specific studies in every

plant. I had set myself once, in a bye-study for foreground drawing, hard on this point; and began, with Mr. Burgess, a complete analysis of the foliation of annual stems; of which Line-studies II., III., and IV., are examples; reduced copies, all, from the beautiful Flora Danica. But after giving two whole lovely long summer days, under the Giesbach, to the blue scabious, ('Devil's bit,') and getting in that time, only half-way up it, I gave in; and must leave the work to happier and younger souls.

25. For these flowering stems, therefore, possessing nearly all the complex organization of a tree, but not its permanence, we will keep the word 'virga;' and 'virgula' for those that have no leaves. I believe, when we come to the study of leaf-order, it will be best to begin with these annual virgæ, in which the leaf has nothing to do with preparation for a next year's branch. And now the remaining terms commonly applied to stems may be for the most part dispensed with; but several are interesting, and must be examined before dismissal.

26. Indeed, in the first place, the word we have to use so often, 'stalk,' has not been got to the roots of, yet. It comes from the Greek στέλεχος, (stelechos,) the 'holding part' of a tree, that which

is like a handle to all its branches; 'stock' is another form in which it has come down to us: with some notion of its being the mother of branches: thus, when Athena's olive was burnt by the Persians, two days after, a shoot a cubit long had sprung from the 'stelechos' of it.

27. Secondly. Few words are more interesting to the modern scholarly and professorial mind than 'stipend.' (I have twice a year at present to consider whether I am worth mine, sent with compliments from the Curators of the University chest.) Now, this word comes from 'stips,' small pay, which itself comes from 'stipo,' to press together, with the idea of small coin heaped up in little towers or piles. But with the idea of lateral pressing together, instead of downward, we get 'stipes,' a solid log; in Greek, with the same sense, στύπος, (stupos,) whence, gradually, with help from another word meaning to beat, (and a side-glance at beating of hemp,) we get our 'stupid,' the German stumph, the Scottish sumph, and the plain English 'stump.'

Refining on the more delicate sound of stipes, the Latins got 'stipula,' the thin stem of straw: which rustles and ripples daintily in verse, associated with spica and spiculum, used of the sharp pointed ear of corn, and its fine processes of fairy shafts.

28. There are yet two more names of stalk to

be studied, though, except for particular plants, not needing to be used,—namely, the Latin cau-dex, and cau-lis, both connected with the Greek καυλός, properly meaning a solid stalk like a handle, passing into the sense of the hilt of a sword, or quill of a pen. Then, in Latin, caudex passes into the sense of log, and so, of cut plank or tablet of wood; thus finally becoming the classical 'codex' of writings engraved on such wooden tablets, and therefore generally used for authoritative manuscripts.

Lastly, 'caulis,' retained accurately in our cauliflower, contracted in 'colewort,' and refined in 'kail,' softens itself into the French 'chou,' meaning properly the whole family of thick-stalked eatable salads with spreading heads; but these being distinguished explicitly by Pliny as 'Capitati,' 'salads with a head,' or 'Captain salads,' the mediæval French softened the 'caulis capitatus' into 'chou cabus;'—or, to separate the round or apple-like mass of leaves from the flowery foam, 'cabus' simply, by us at last enriched and emphasized into 'cabbage.'

29. I believe we have now got through the stiffest piece of etymology we shall have to master in the course of our botany; but I am certain that young readers will find patient work, in this kind, well rewarded by the groups of connected thoughts

which will thus attach themselves to familiar names; and their grasp of every language they learn must only be esteemed by them secure when they recognize its derivatives in these homely associations, and are as much at ease with the Latin or French syllables of a word as with the English ones; this familiarity being above all things needful to cure our young students of their present ludicrous impression that what is simple, in English, is knowing, in Greek; and that terms constructed out of a dead language will explain difficulties which remained insoluble in a living one. But Greek is *not* yet dead: while if we carry our unscholarly nomenclature much further, English soon will be; and then doubtless botanical gentlemen at Athens will for some time think it fine to describe what we used to call caryophyllaceæ, as the ἐδληφιδες.

30. For indeed we are all of us yet but schoolboys, clumsily using alike our lips and brains; and with all our mastery of instruments and patience of attention, but few have reached, and those dimly, the first level of science,—wonder.

For the first instinct of the stem,—unnamed by us yet—unthought of,—the instinct of seeking light, as of the root to seek darkness,—what words can enough speak the wonder of it!

Look. Here is the little thing, Line-study V.

(A), in its first birth to us: the stem of stems; the one of which we pray that it may bear our daily bread. The seed has fallen in the ground with the springing germ of it downwards; with heavenly cunning the taught stem curls round, and seeks the never-seen light. Veritable 'conversion,' miraculous, called of God. And here is the oat germ, (B)—after the wheat, most vital of divine gifts; and assuredly, in days to come, fated to grow on many a naked rock in hitherto lifeless lands, over which the glancing sheaves of it will shake sweet treasure of innocent gold.

And who shall tell us how they grow; and the fashion of their rustling pillars—bent, and again erect, at every breeze. Fluted shaft or clustered pier, how poor of art, beside this grass-shaft—built, first to sustain the food of men, then to be strewn under their feet!

We must not stay to think of it, *yet*, or we shall get no farther till harvest has come and gone again. And having our names of stems now determined enough, we must in next chapter try a little to understand the different kinds of them.

Line-Study II.

Line-Study III.

Line-Study IV.

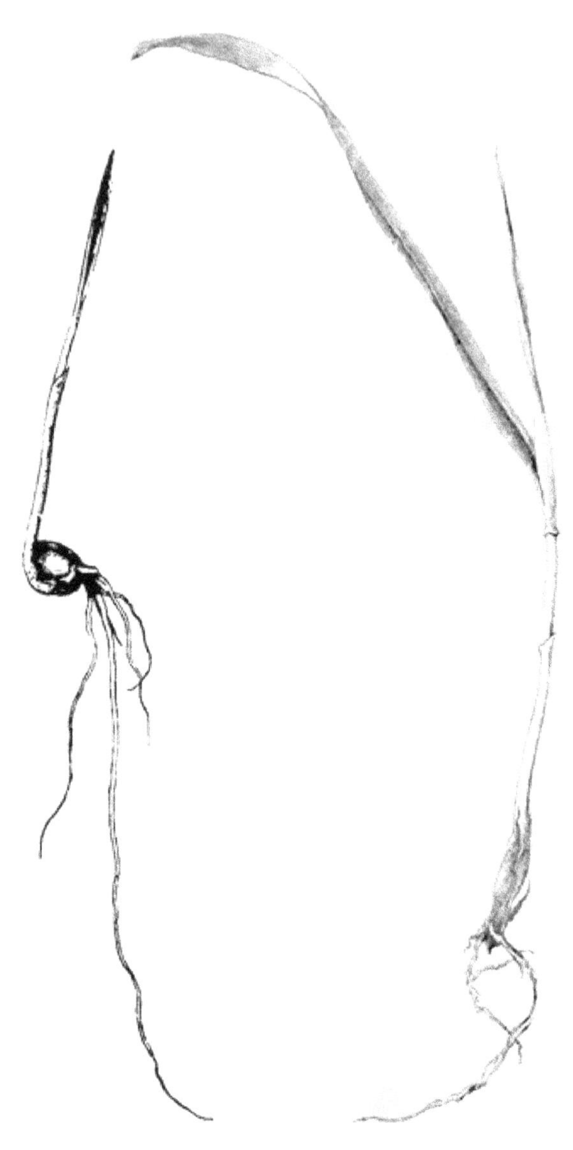

Line-Study V.

"Behold, a Sower went forth to sow."

The following notes, among many kindly sent me on the subject of Scottish Heraldry, seem to be the most trustworthy:—

"The earliest known mention of the thistle as the national badge of Scotland is in the inventory of the effects of James III.; who probably adopted it as an appropriate illustration of the royal motto, *In defence.*

"Thistles occur on the coins of James IV., Mary, James V., and James VI.; and on those of James VI. they are for the first time accompanied by the motto, *Nemo me impune lacesset.*

"A collar of thistles appears on the gold bonnet-pieces of James V. of 1539; and the royal ensigns, as depicted in Sir David Lindsay's armorial register of 1542, are surrounded by a collar formed entirely of golden thistles, with an oval badge attached.

"This collar, however, was a mere device until the institution, or, as it is generally but inaccurately called, the revival, of the order of the Thistle by James VII. (II. of England), which took place on May 29, 1687."

Date of James III.'s reign 1460—1488.

CHAPTER IX.

OUTSIDE AND IN.

1. THE elementary study of methods of growth, given in the following chapter, has been many years written, (the greater part soon after the fourth volume of 'Modern Painters'); and ought now to be rewritten entirely; but having no time to do this, I leave it with only a word or two of modification, because some truth and clearness of incipient notion will be conveyed by it to young readers, from which I can afterwards lop the errors, and into which I can graft the finer facts, better than if I had a less blunt embryo to begin with.

2. A stem, then, broadly speaking, (I had thus began the old chapter,) is the channel of communication between the leaf and root; and if the leaf can grow directly from the root, there is no stem: so that it is well first to conceive of all plants as consisting of leaves and roots only, with the condition that each leaf must have its own

quite particular root * somewhere. Let a b c, Fig. 16, be three leaves, each, as you see, with its own root, and by no means dependent on other leaves for its daily bread; and let the horizontal line be the surface of the ground. Then the plant has no stem, or an underground one. But if the three leaves rise above the ground, as in Fig. 17, they must reach their roots by elongating their stalks, and this elongation is the stem of the plant. If the outside leaves grow last and are therefore youngest, the plant is said to grow from the outside. You know that 'ex' means out, and that 'gen' is the first syllable of Genesis (or creation), therefore the old botanists, putting an o between the two syllables, called plants whose outside leaves grew last, Ex-o-gens. If the inside leaf grows last, and is youngest, the plant was said to grow from the inside, and from the Greek Endon, within, called an 'Endo-gen.' If

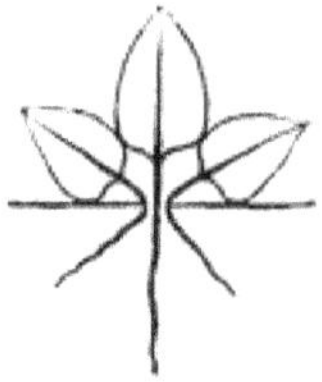

Fig. 16.

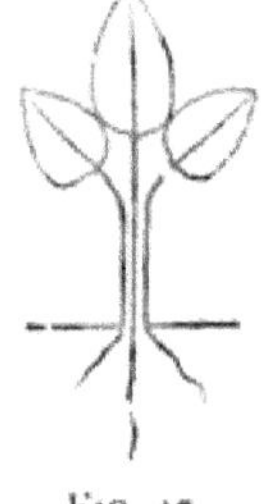

Fig. 17.

* Recent botanical research makes this statement more than dubitable. Nevertheless, on no other supposition can the forms and action of tree-branches, so far as at present known to me, be yet clearly accounted for.

these names are persisted in, the Greek botanists, to return the compliment, will of course call Endogens Ἰνσειδβορνιδες, and Exogens Ὀυτσειδβορνιδες. In the Oxford school, they will be called simply Inlaid and Outlaid.

3. You see that if the outside leaves are to grow last, they may conveniently grow two at a time; which they accordingly do, and exogens always start with two little leaves from their roots, and may therefore conveniently be called two-leaved; which, if you please, we will for our parts call them. The botanists call them 'two-suckered,' and can't be content to call them *that* in English; but drag in a long Greek word, meaning the fleshy sucker of the sea-devil,—'cotyledon,' which, however, I find is practically getting shortened into 'cot,' and that they will have to end by calling endogens, monocots, and exogens, bicots. I mean steadily to call them one-leaved and two-leaved, for this further reason, that they differ not merely in the single or dual springing of first leaves from the seed; but in the distinctly single or dual arrangement of leaves afterwards on the stem; so that, through all the complexity obtained by alternate and spiral placing, every bicot or two-leaved flower or tree is in reality composed of dual groups of leaves, separated by a given length of stem; as, most characteristically in

this pure mountain type of the Ragged Robin (Clarissa laciniosa), Fig. 18; and compare A, and B, Line-study II.; while, on the other hand, the monocot plants are by close analysis, I think, always resolvable into successively climbing leaves, sessile on one another, and sending their roots, or processes, for nourishment, down through one another, as in Fig. 19.

4. Not that I am yet clear, at all, myself; but I do think it's more the botanists' fault than mine, what 'cotyledonous' structure there may be at the outer base of each successive bud; and still less, how the intervenient length of stem, in the bicots, is related to their power, or law, of branching. For not only the two-leaved tree is outlaid, and the one-leaved inlaid, but the two-leaved tree is branched, and the one-leaved tree is not branched. This is a most vital and important distinction, which I state to you in very bold terms, for

Fig. 18.

Fig. 19.

though there are some apparent exceptions to the law, there are, I believe, no real ones, if we define a branch rightly. Thus, the head of a palm tree is merely a cluster of large leaves; and the spike of a grass, a clustered blossom. The stem, in both, is unbranched; and we should be able in this respect to classify plants very simply indeed, but for a provoking species of intermediate creatures whose branching is always in the manner of corals, or sponges, or aborescent minerals, irregular and accidental, and essentially, therefore, distinguished from the systematic anatomy of a truly branched tree. Of these presently; we must go on by very short steps: and I find no step can be taken without check from existing generalizations. Sowerby's definition of Monocotyledons, in his ninth volume, begins thus: "Herbs, (or rarely, and only in exotic genera,) trees, in which the wood, pith, and bark are indistinguishable." Now if there be one plant more than another in which the pith is defined, it is the common Rush; while the nobler families of true herbs derive their principal character from being pithless altogether! We cannot advance too slowly.

5. In the families of one-leaved plants in which the young leaves grow directly out of the old ones,

it becomes a grave question for them whether the old ones are to lie flat or edgeways, and whether they must therefore grow out of their faces or their edges. And we must at once understand the way they contrive it, in either case.

Among the many forms taken by the Arethusan leaf, one of the commonest is long and gradually tapering,—much broader at the base than the point. We will take such an one for examination, and suppose that it is growing on the ground as in Fig. 20, with a root to its every fibre. Cut out a piece of strong paper roughly into the shape of this Arethusan leaf, a, Fig. 21. Now suppose the next young leaf has to spring out of the front of this one, at about the middle of its height. Give it two nicks with the scissors at b b; then roll up the lower part into a cylinder, (it will overlap a good deal at the bottom,) and tie it fast with a fine thread: so, you will get the form at c. Then bend the top of it back, so that, seen sideways, it appears as at d, and you see you have made quite a little flower-pot to plant your new leaf in, and perhaps it may occur to you that you have seen

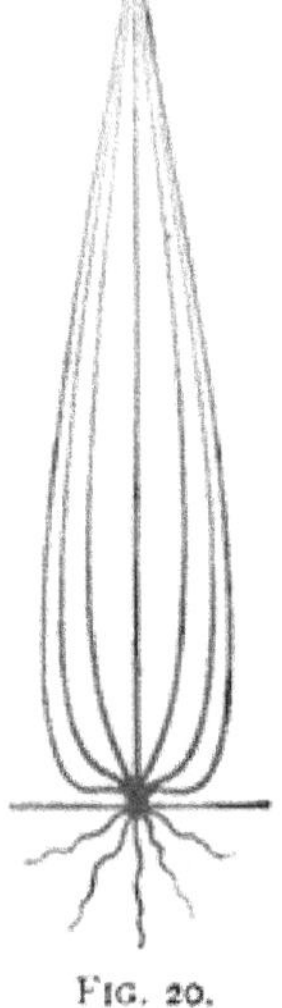
FIG. 20.

something like this before. Now make another, a little less wide, but with the part for the cylinder twice as long, roll it up in the same way, and slip it inside the other, with the flat part turned the

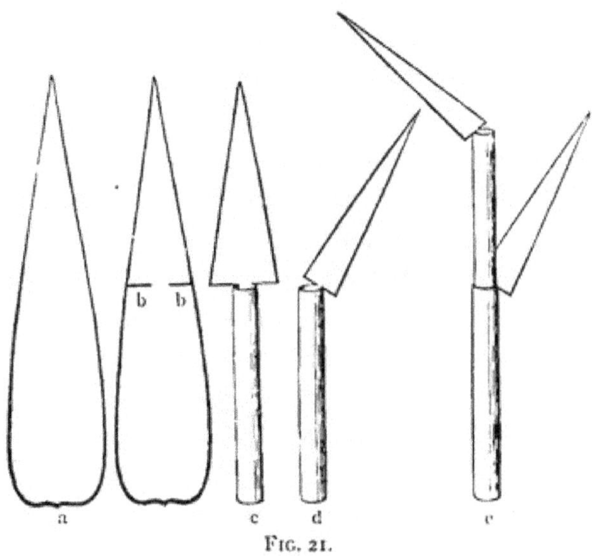

FIG. 21.

other way, e. Surely this reminds you now of something you have seen? Or must I draw the something (Fig. 22)?

6. All grasses are thus constructed, and have their leaves set thus, opposite, on the sides of their tubular stems, alternately, as they ascend. But in most of them there is also a peculiar construction, by which, at the base of the sheath, or enclosing

tube, each leaf articulates itself with the rest of the stem at a ringed knot, or joint.

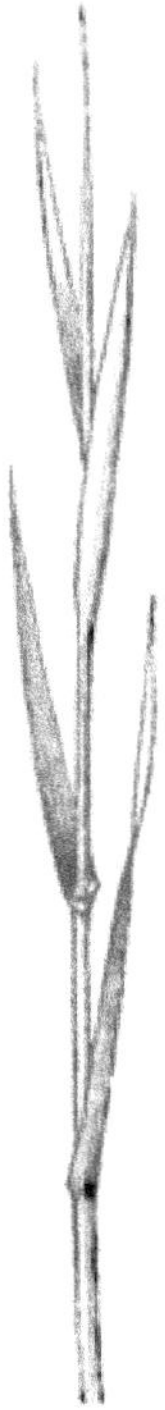

Before examining these, remember there are mainly two sorts of joints in the framework of the bodies of animals. One is that in which the bone is thick at the joints and thin between them, (see the bone of the next chicken leg you eat,) the other is that of animals that have shells or horny coats, in which characteristically the shell is thin at the joints, and thick between them (look at the next lobster's claw you can see, without eating). You know, also, that though the crustaceous are titled only from their crusts, the name 'insect' is given to the whole insect tribe, because they are farther jointed almost into *sections*; it is easily remembered, also, that the projecting joint means strength and elasticity in the creature, and that all its limbs are useful to it, and cannot conveniently be parted with; and that the incised, sectional, or insectile joint means more or less weakness,* and necklace-like laxity or license in the creature's make; and an ignoble power

FIG. 22.

* Not always in muscular power; but the framework on which strong muscles are to act, as that of an insect's wing, or its jaw, is never insectile.

of shaking off its legs or arms on occasion, coupled also with modes of growth involving occasionally quite astonishing transformations, and beginnings of new life under new circumstances; so that, until very lately, no mortal knew what a crab was like in its youth, the very existence of the creature, as well as its legs, being jointed, as it were, and made in separate pieces with the narrowest possible thread of connection between them; and its principal, or stomachic, period of life, connected with its sentimental period by as thin a thread as a wasp's stomach is with its thorax.

7. Now in plants, as in animals, there are just the same opposed aspects of joint, with this specialty of difference in function, that the animal's limb bends at the joints, but the vegetable limb stiffens. And when the articulation projects, as in the joint of a cane, it means not only that the strength of the plant is well carried through the junction, but is carried farther and more safely than it could be without it: a cane is stronger, and can stand higher that it could otherwise, because of its joints. Also, this structure implies that the plant has a will of its own, and a position which on the whole it will keep, however it may now and then be bent out of it; and that it has a continual battle, of a healthy and human-like kind, to wage with surrounding elements.

But the crabby, or insect-like, joint, which you get

in seaweeds and cacti, means either that the plant is to be dragged and wagged here and there at the will of waves, and to have no spring nor mind of its own ; or else that it has at least no springy intention and elasticity of purpose, but only a knobby, knotty, prickly, malignant stubbornness, and incoherent opiniativeness ; crawling about, and coggling, and grovelling, and aggregating anyhow, like the minds of so many people whom one knows!

8. Returning then to our grasses, in which the real rooting and junction of the leaves with each other is at these joints ; we find that therefore every leaf of grass may be thought of as consisting of two main parts, for which we shall want two separate names. The lowest part, which wraps itself round to become strong, we will call the 'staff,' and for the free-floating outer part we will take specially the name given at present carelessly to a large number of the plants themselves, 'flag.' This will give a more clear meaning to the words 'rod' (virga), and 'staff' (baculus), when they occur together, as in the 23rd Psalm ; and remember the distinction is that a rod bends like a switch, but a staff is stiff. I keep the well-known name 'blade' for grass-leaves in their fresh green state.

9. You felt, as you were bending down the paper into the form d, Fig. 21, the difficulty and

awkwardness of the transition from the tubular form of the staff to the flat one of the flag. The mode in which this change is effected is one of the most interesting features in plants, for you will find presently that the leaf-stalk in ordinary leaves is only a means of accomplishing the same change from round to flat. But you know I said just now that some leaves were not flat, but set upright, edgeways. It is not a common position in two-leaved trees; but if you can run out and look at an arbor vitæ, it may interest you to see its hatchet-shaped vertically crested cluster of leaves transforming themselves gradually downwards into branches; and in one-leaved trees the vertically edged group is of great importance.

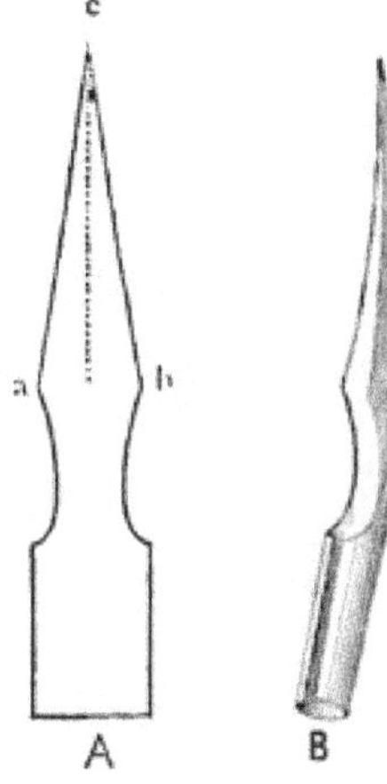

Fig. 23.

10. Cut out another piece of paper like a in Fig. 21, but now, instead of merely giving it nicks at a, b, cut it into the shape A, Fig. 23. Roll the lower part up as before, but instead of pulling the upper part down, pinch its back at the dotted line, and bring the two points, a and b, forward, so that they may touch each other. B shows the look of the

thing half-done, before the points a and b have quite met. Pinch them close, and stitch the two edges neatly together, all the way from a to the point c; then roll and tie up the lower part as before. You will find then that the back or spinal line of the whole leaf is bent forward, as at B. Now go out to the garden and gather the green leaf of a fleur-de-lys, and look at it and your piece of disciplined paper together; and I fancy you will probably find out several things for yourself that I want you to know.

11. You see, for one thing, at once, how *strong* the fleur-de-lys leaf is, and that it is just twice as strong as a blade of grass, for it is the substance of the staff, with its sides flattened together, while the grass blade is a staff cut open and flattened out. And you see that as a grass blade necessarily flaps down, the fleur-de-lys leaf as necessarily curves up, owing to that inevitable bend in its back. And you see, with its keen edge, and long curve, and sharp point, how like a sword it is. The botanists would for once have given a really good and right name to the plants which have this kind of leaf, 'Ensatæ,' from the Latin 'ensis,' a sword; if only sata had been properly formed from sis. We can't let the rude Latin stand, but you may remember that the fleur-de-lys, which is the flower of chivalry, has a sword for its leaf, and a lily for its heart.

12. In case you cannot gather a fleur-de-lys leaf, I have drawn for you, in Plate VI., a cluster of such leaves, which are as pretty as any, and so small that, missing the points of a few, I can draw them of their actual size. You see the pretty alternate interlacing at the bottom, and if you can draw at all, and will try to outline their curves, you will find what subtle lines they are. I did not know this name for the strong-edged grass leaves when I wrote the pieces about shield and sword leaves in 'Modern Painters'; I wish I had chanced in those passages on some other similitude, but I can't alter them now, and my trustful pupils may avoid all confusion of thought by putting gladius for ensis, and translating it by the word 'scymitar,' which is also more accurate in expressing the curvature blade. So we will call the ensatæ, instead, 'gladiolæ,' translating, 'scymitar-grasses.' And having now got at some clear idea of the distinction between outlaid and inlaid growth in the stem, the reader will find the elementary analysis of forms resulting from outlaid growth in 'Modern Painters'; and I mean to republish it in the sequel of this book, but must go on to other matters here. The growth of the inlaid stem we will follow as far as we need, for English plants, in examining the grasses.

Drawn by J Ruskin — Engraved by G Allen

VI

Radical Insertion of leaves of Ensatæ.

IRIS GERMANICA.

FLORENCE, 11*th September*, 1874.

As I correct this chapter for press, I find it is too imperfect to be let go without a word or two more. In the first place, I have not enough, in distinguishing the nature of the living yearly shoot, with its cluster of fresh leafage, from that of the accumulated mass of perennial trees, taken notice of the similar power even of the annual shoot, to obtain some manner of immortality for itself, or at least of usefulness, *after* death. A Tuscan woman stopped me on the path up to Fiesole last night, to beg me to buy her plaited straw. I wonder how long straw lasts, if one takes care of it? A Leghorn bonnet, (if now such things are,) carefully put away,—even properly taken care of when it is worn,—how long will it last, young ladies?

I have just been reading the fifth chapter of II. Esdras, and am fain to say, with less discomfort than otherwise I might have felt, (the example being set me by the archangel Uriel,) "I am not sent to tell thee, for I do not know." How old is the oldest straw known? the oldest linen? the oldest hemp? We have mummy wheat,—cloth of papyrus, which is a kind of straw. The paper reeds by the brooks, the flax-flower in the field, leave such imperishable frame behind them. And Ponte-della-Paglia, in Venice; and Straw Street, of

Paris, remembered in Heaven,—there is no occasion to change their names, as one may have to change 'Waterloo Bridge,' or the 'Rue de l'Impératrice.' Poor Empress! Had she but known that her true dominion was in the straw streets of her fields; not in the stone streets of her cities!

But think how wonderful this imperishableness of the stem of many plants is, even in their annual work: how much more in their perennial work! The noble stability between death and life, of a piece of perfect wood? It cannot grow, but will not decay; keeps record of its years of life, but surrenders them to become a constantly serviceable thing: which may be sailed in, on the sea, built with, on the land, carved by Donatello, painted on by Fra Angelico. And it is not the wood's fault, but the fault of Florence in not taking proper care of it, that the panel of Sandro Botticelli's loveliest picture has cracked, (not with heat, I believe, but blighting frost,) a quarter of an inch wide through the Madonna's face.

But what is this strange state of undecaying wood? What sort of latent life has it, which it only finally parts with when it rots?

Nay, what is the law by which its natural life is measured? What makes a tree 'old'? One sees the Spanish-chestnut trunks among the Apennines growing into caves, instead of logs. Vast hollows, confused among the recessed darknesses of the marble crags, surrounded by mere laths of living stem, each with its coronal of glorious green leaves. Why can't the tree go on, and on,—hollowing itself into a Fairy—no—a Dryad, Ring,—till it becomes a perfect Stonehenge of a tree? Truly, "I am not sent to tell thee, for I do not know."

The worst of it is, however, that I don't know one thing which I ought very thoroughly to have known at least thirty years ago, namely, the true difference in the way of building the trunk in outlaid and inlaid wood. I have an idea that the stem of a palm-tree is only a heap of leaf-roots built up like a tower of bricks, year by year, and that the palm-tree really grows on the top of it, like a bunch of fern; but I've no books here, and no time to read them if I had. If only I were a strong giant, instead of a thin old gentleman of fifty-five, how I should like to pull up one of those little palm-trees by the roots—(by the way, what are the roots of a palm like? and, how does it stand in sand, where it is wanted to stand, mostly? Fancy, not knowing that, at fifty-five!) —that grow all along the Riviera; and snap its stem in two, and cut it down the middle. But I suppose there are sections enough now in our grand botanical collections, and you can find it all out for yourself. That you should be able to ask a question clearly, is two-thirds of the way to getting it answered; and I think this chapter of mine will at least enable you to ask some questions about the stem, though what a stem *is*, truly, "I am not sent to tell thee, for I do not know."

KNARESBOROUGH, 30*th April*, 1876.

I see by the date of last paragraph that this chapter has been in my good Aylesbury printer's type for more than a year and a half. At this rate, Proserpina has a distant chance of being finished in the spirit-land, with more accurate information derived from the archangel Uriel himself, (not that he is likely to know much about the matter, if he keeps on letting himself be prevented from ever seeing foliage in spring-time by the black demon-winds,) about the year 2000. In the meantime feeling that perhaps I *am* sent to tell my readers a little more than is above told, I have had recourse to my botanical friend, good Mr. Oliver of Kew, who has taught me, first, of palms, that they actually stitch themselves into the ground, with a long dipping loop, up and down, of the root fibres, concerning which sempstress-work I shall have a month's puzzlement before I can report on it; secondly, that all the increment of tree stem is, by division and multiplication of the cells of the wood, a process not in the least to be described as 'sending down roots from the leaf to the ground.' I suspected as much in beginning to revise this chapter; but hold to my judgment in not cancelling it. For this multiplication of the cells is at least compelled by an influence which passes from the leaf to the ground, and vice versâ; and which is at present best conceivable to me by imagining the continual and invisible descent of lightning from electric cloud by a conducting rod, endowed with the power of softly splitting the rod into two rods, each as thick as the original one. Studying microscopically, we should then see the molecules of copper, as we see the cells of the wood, dividing and increasing, each one of them into two. But the visible result, and mechanical conditions of growth, would still be the same as if the leaf actually sent down a new root fibre; and, more than this, the currents of accumulating substance, marked by the grain of the wood, are, I think, quite plainly and absolutely those of streams flowing only from the leaves downwards; never from the root up, nor of mere lateral increase. I must look over all my drawings again, and at tree stems again, with more separate study of the bark and pith in those museum sections, before I can assert this; but there will be no real difficulty in the investigation. If the increase of the wood is lateral only, the currents round the knots will be compressed at the sides, and open above and below; but if downwards, compressed above the knot and open below it. The nature of the force itself, and the manner of its ordinances in direction, remain, and must for ever remain, inscrutable as our own passions, in the hand of the God of all Spirits, and of all Flesh.

"Drunk is each ridge, of thy cup drinking,
Each clod relenteth at thy dressing,
Thy cloud-borne waters inly sinking,
Fair spring sproutes forth, blest with thy blessing;
The fertile year is with thy bounty crouned,
And where thou go'st, thy goings fat the ground.

Plenty bedews the desert places,
A hedge of mirth the hills encloseth.
The fields with flockes have hid their faces,
A robe of corn the valleys clotheth.
Deserts and hills and fields and valleys all,
Rejoice, shout, sing, and on thy name do call."

CHAPTER X.

THE BARK.

1. PHILOLOGISTS are continually collecting instances, like our friend the French critic of Virgil, of the beauty of finished language, or the origin of unfinished, in the imitation of natural sounds. But such collections give an entirely false idea of the real power of language, unless they are balanced by an opponent list of the words which signally fail of any such imitative virtue, and whose sound, if one dwelt upon it, is destructive of their meaning.

2. For instance. Few sounds are more distinct in their kind, or one would think more likely to be vocally reproduced in the word which signified them, than that of a swift rent in strongly woven cloth; and the English words 'rag' and ragged, with the Greek *ῥήγνυμι*, do indeed in a measure recall the tormenting effect upon the ear. But it is curious that the verb which is meant to express the actual origination of rags, should

rhyme with two words entirely musical and peaceful—words, indeed, which I always reserve for final resource in passages which I want to be soothing as well as pretty,—'fair,' and 'air'; while, in its orthography, it is identical with the word representing the bodily sign of tenderest passion, and grouped with a multitude of others,* in which the mere insertion of a consonant makes such wide difference of sentiment as between 'dear' and 'drear,' or 'pear' and 'spear.' The Greek root, on the other hand, has persisted in retaining some vestige of its excellent dissonance, even where it has parted with the last vestige of the idea it was meant to convey; and when Burns did his best,—and his best was above most men's,—to gather pleasant liquid and labial syllabling round gentle meaning, in

"Bonnie lassie, will ye go,
Will ye go, will ye go,
Bonnie lassie, will ye go,
To the birks of Aberfeldy?"

he certainly had little thought that the delicately crisp final k, in birk, was the remnant of a magnificent Greek effort to express the rending of the

* It is one of the three cadences, (the others being of the words rhyming to 'mind' and 'way,') used by Sir Philip Sidney in his marvellous paraphrase of the 55th Psalm.

earth by earthquake, in the wars of the giants. In the middle of that word 'esmaragēse,' we get our own beggar's 'rag' for a pure root, which afterwards, through the Latin frango, softens into our 'break,' and 'bark,'—the 'broken thing'; that idea of its rending around the tree's stem having been, in the very earliest human efforts at botanical description, attached to it by the pure Aryan race, watching the strips of rosy satin break from the birch stems, in the Aberfeldys of Imaus.

3. That this tree should have been the only one which "the Aryans, coming as conquerors from the North, were able to recognize in Hindostan,"* and should therefore also be "the only one whose name is common to Sanskrit, and to the languages of Europe," delighted me greatly, for two reasons: the first, for its proof that in spite of the development of species, the sweet gleaming of birch stem has never changed its argent and sable for any unchequered heraldry; and the second, that it gave proof of a much more important fact, the keenly accurate observation of Aryan foresters at that early date; for the fact is that the breaking of the thin-beaten silver of the birch trunk is so delicate, and its smoothness so graceful, that until I painted it with

* Lectures on the Families of Speech, by the Rev. F. Farrer. Longman, 1870. Page 81.

care, I was not altogether clear-headed myself about the way in which the chequering was done: nor until Fors to-day brought me to the house of one of my father's friends at Carshalton, and gave me three birch stems to look at just outside the window, did I perceive it to be a primal question about them, what it is that blanches that dainty dress of theirs, or, anticipatorily, weaves. What difference is there between the making of the corky excrescence of other trees, and of this almost transparent fine white linen? I perceive that the older it is, within limits, the finer and whiter; hoary tissue, instead of hoary air—honouring the tree's aged body; the outer sprays have no silvery light on their youth. Does the membrane thin itself into whiteness merely by stretching, or produce an outer film of new substance?*

4. And secondly, this investiture, why is it transverse to the trunk,—swathing it, as it were, in bands? Above all,—when it breaks,—why does it break round the tree instead of down? All other bark breaks as anything would, naturally, round a swelling rod, but this, as if the stem were growing longer; until, indeed, it reaches farthest heroic old

* I only profess, you will please to observe, to ask questions in Proserpina. Never to answer any. But of course this chapter is to introduce some further inquiry in another place.

age, when the whiteness passes away again, and the rending is like that of other trees, downwards. So that, as it were in a changing language, we have the great botanical fact twice taught us, by this tree of Eden, that the skins of trees differ from the skins of the higher animals in that, for the most part, they won't stretch, and must be worn torn.

So that in fact the most popular arrangement of vegetative adult costume is Irish; a normal investiture in honourable rags; and decorousness of tattering, as of a banner borne in splendid ruin through storms of war.

5. Now therefore, if we think of it, we have five distinct orders of investiture for organic creatures; first, mere secretion of mineral substance, chiefly lime, into a hard shell, which, if broken, can only be mended, like china—by sticking it together; secondly, organic substance of armour which grows into its proper shape at once for good and all, and can't be mended at all, if broken, (as of insects); thirdly, organic substance of skin, which stretches, as the creature grows, by cracking, over a fresh skin which is supplied beneath it, as in bark of trees; fourthly, organic substance of skin cracked symmetrically into plates or scales which can increase all round their edges, and are

connected by softer skin, below, as in fish and reptiles, (divided with exquisite lustre and flexibility, in feathers of birds); and lastly, true elastic skin, extended in soft unison with the creature's growth, —blushing with its blood, fading with its fear; breathing with its breath, and guarding its life with sentinel beneficence of pain.

6. It is notable, in this higher and lower range of organic beauty, that the decoration, by pattern and colour, which is almost universal in the protective coverings of the middle ranks of animals, should be reserved in vegetables for the most living part of them, the flower only: and that among animals, few but the malignant and senseless are permitted, in the corrugation of their armour, to resemble the half-dead trunk of the tree, as they float beside it in the tropical river. I must, however, leave the scale patterns of the palms and other inlaid tropical stems for after-examination,—content, at present, with the general idea of the bark of an outlaid tree as the successive accumulation of the annual protecting film, rent into ravines of slowly increasing depth, and coloured, like the rock, whose stability it begins to emulate, with the grey or gold of clinging lichen and embroidering moss.

CHAPTER XI.

GENEALOGY.

1. RETURNING, after more than a year's sorrowful interval, to my Sicilian fields,—not incognisant, now, of some of the darker realms of Proserpina; and with feebler heart, and, it may be, feebler wits, for wandering in her brighter ones,—I find what I had written by way of sequel to the last chapter, somewhat difficult, and extremely tiresome. Not the less, after giving fair notice of the difficulty, and asking due pardon for the tiresomeness, I am minded to let it stand; trusting to end, with it, once for all, investigations of the kind. But in finishing this first volume of my School Botany, I must try to give the reader some notion of the plan of the book, as it now, during the time for thinking over it which illness left me, has got itself arranged in my mind, within limits of possible execution. And this the rather, because I wish also to state, somewhat more gravely than I have yet done, the

grounds on which I venture here to reject many of the received names of plants; and to substitute others for them, relating to entirely different attributes from those on which their present nomenclature is confusedly edified.

I have already in some measure given the reasons for this change;* but I feel that, for the sake of those among my scholars who have laboriously learned the accepted names, I ought now also to explain its method more completely.

2. I call the present system of nomenclature *confusedly* edified, because it introduces,—without, apparently, any consciousness of the inconsistency, and certainly with no apology for it,—names founded sometimes on the history of plants, sometimes on their qualities, sometimes on their forms, sometimes on their products, and sometimes on their poetical associations.

On their history—as 'Gentian' from King Gentius, and 'Funkia' from Dr. Funk.

On their qualities—as 'Scrophularia' from its (quite uncertified) use in scrofula.

On their forms—as the 'Caryophylls' from having petals like husks of nuts.

On their products—as 'Cocos nucifera' from its nuts.

* See Introduction, pp. 5—8.

And on their poetical associations,—as the 'Star of Bethlehem' from its imagined resemblance to the light of that seen by the Magi.

3. Now, this variety of grounds for nomenclature might patiently, and even with advantage, be permitted, provided the grounds themselves were separately firm, and the inconsistency of method advisedly allowed, and, in each case, justified. If the histories of King Gentius and Dr. Funk are indeed important branches of human knowledge;—if the Scrophulariaceæ do indeed cure King's Evil;—if pinks be best described in their likeness to nuts;—and the Star of Bethlehem verily remind us of Christ's Nativity,—by all means let these and other such names be evermore retained. But if Dr. Funk be not a person in any special manner needing either stellification or florification; if neither herb nor flower can avail, more than the touch of monarchs, against hereditary pain; if it be no better account of a pink to say it is nut-leaved, than of a nut to say it is pink-leaved; and if the modern mind, incurious respecting the journeys of wise men, has already confused, in its Bradshaw's Bible, the station of Bethlehem with that of Bethel,* it is certainly time to take some order with the partly false, partly useless, and partly

* See Sowerby's nomenclature of the flower, vol. ix., plate 1703.

forgotten literature of the Fields; and, before we bow our children's memories to the burden of it, ensure that there shall be matter worth carriage in the load.

4. And farther, in attempting such a change, we must be clear in our own minds whether we wish our nomenclature to tell us something about the plant itself, or only to tell us the place it holds in relation to other plants: as, for instance, in the Herb-Robert, would it be well to christen it, shortly, 'Rob Roy,' because it is pre-eminently red, and so have done with it;—or rather to dwell on its family connections, and call it 'Macgregoraceous'?

5. Before we can wisely decide this point, we must resolve whether our botany is intended mainly to be useful to the vulgar, or satisfactory to the scientific élite. For if we give names characterizing individuals, the circle of plants which any country possesses may be easily made known to the children who live in it: but if we give names founded on the connexion between these and others at the Antipodes, the parish schoolmaster will certainly have double work; and it may be doubted greatly whether the parish schoolboy, at the end of the lecture, will have half as many ideas.

6. Nevertheless, when the features of any great order of plants are constant, and, on the whole,

VII.

CONTORTA PURPUREA.

PURPLE WREATH-WORT.

represented with great clearness both in cold and warm climates, it may be desirable to express this their citizenship of the world in definite nomenclature. But my own method, so far as hitherto developed, consists essentially in fastening the thoughts of the pupil on the special character of the plant, in the place where he is likely to see it; and therefore, in expressing the power of its race and order in the wider world, rather by reference to mythological associations than to botanical structure.

7. For instance, Plate VII. represents, of its real size, an ordinary spring flower in our English mountain fields. It is an average example,—not one of rare size under rare conditions,—rather smaller than the average, indeed, that I might get it well into my plate. It is one of the flowers whose names I think good to change; but I look carefully through the existing titles belonging to it and its fellows, that I may keep all I expediently can. I find, in the first place, that Linnæus called one group of its relations, Ophryds, from Ophrys,—Greek for the eyebrow,—on account of their resemblance to the brow of an animal frowning, or to the overshadowing casque of a helmet. I perceive this to be really a very general aspect of the flower; and therefore, no less than in respect

to Linnæus, I adopt this for the total name of the order, and call them 'Ophrydæ,' or, shortly, 'Ophryds.'

8. Secondly: so far as I know these flowers myself, I perceive them to fall practically into three divisions,—one, growing in English meadows and Alpine pastures, and always adding to their beauty; another, growing in all sorts of places, very ugly itself, and adding to the ugliness of its indiscriminated haunts; and a third, growing mostly up in the air, with as little root as possible, and of gracefully fantastic forms, such as this kind of nativity and habitation might presuppose. For the present, I am satisfied to give names to these three groups only. There may be plenty of others which I do not know, and which other people may name, according to their knowledge. But in all these three kinds known to me, I perceive one constant characteristic to be *some* manner of *distortion*; and I desire that fact,—marking a spiritual (in my sense of the word) character of extreme mystery,—to be the first enforced on the mind of the young learner. It is exhibited to the English child, primarily, in the form of the stalk of each flower, attaching it to the central virga. This stalk is always twisted once and a half round, as if some-

body had been trying to wring the blossom off; and the name of the family, in Proserpina, will therefore be 'Contorta'* in Latin, and 'Wreathe-wort' in English.

Farther: the beautiful power of the one I have drawn in its spring life, is in the opposition of its dark purple to the primrose in England, and the pale yellow anemone in the Alps. And its individual name will be, therefore, 'Contorta purpurea'—*Purple* Wreathe-wort.

And in drawing it, I take care to dwell on the strength of its colour, and to show thoroughly that it is a *dark* blossom,† before I trouble myself about its minor characters.

9. The second group of this kind of flowers live, as I said, in all sorts of places; but mostly, I think, in disagreeable ones,—torn and irregular ground, under alternations of unwholesome heat and shade, and among swarms of nasty insects. I cannot yet venture on any bold general statement about them, but I think that is mostly their way; and at all events, they themselves are in the habit of dressing in livid and unpleasant

* Linnæus used this term for the Oleanders; but evidently with less accuracy than usual.

† "ἄνθη πορφυροειδῆ" says Dioscorides, of the race generally,—but "ἄνθη δὲ ὑποπόρφυρα" of this particular one.

colours; and are distinguished from all other flowers by twisting, not only their stalks, but one of their petals, not once and a half only, but two or three times round, and putting it far out at the same time, as a foul jester would put out his tongue: while also the singular power of grotesque mimicry, which, though strong also in the other groups of their race, seems in the others more or less playful, is, in these, definitely degraded, and, in aspect, malicious.

10. Now I find the Latin name 'Satyrium' attached already to one sort of these flowers; and we cannot possibly have a better one for all of them. It is true that, in its first Greek form, Dioscorides attaches it to a white, not a livid, flower; and I dare say there are some white ones of the breed: but, in its full sense, the term is exactly right for the entire group of ugly blossoms of which the characteristic is the spiral curve and protraction of their central petal: and every other form of Satyric ugliness which I find among the Ophryds, whatever its colour, will be grouped with them. And I make them central, because this humour runs through the whole order, and is, indeed, their distinguishing sign.

11. Then the third group, living actually in the air, and only holding fast by, without nourishing

itself from, the ground, rock, or tree-trunk on which it is rooted, may of course most naturally and accurately be called 'Aeria,' as it has long been popularly known in English by the name of Air-plant.

Thus we have one general name for all these creatures, 'Ophryd'; and three family or group names, Contorta, Satyrium, and Aeria,—every one of these titles containing as much accurate fact about the thing named as I can possibly get packed into their syllables: and I will trouble my young readers with no more divisions of the order. And if their parents, tutors, or governors, after this fair warning, choose to make them learn, instead, the seventy-seven different names with which botanist-heraldries have beautifully ennobled the family,—all I can say is, let them at least begin by learning them themselves. They will be found in due order in pages 1084, 1085 of Loudon's Cyclopædia.*

12. But now, farther: the student will observe that the name of the total order is Greek; while

* I offer a sample of two dozen for good papas and mammas to begin with:—

Angraecum.	Corallorrhiza.	Ornithidium.	Prescotia.
Anisopetalum.	Cryptarrhena.	Ornithocephalus.	Renanthera.
Brassavola.	Eulophia.	Platanthera.	Rodrigueria.
Brassia.	Gymnadenia.	Pleurothallis.	Stenorhyncus.
Caelogyne.	Microstylis.	Pogonia.	Trizeuxis.
Calopogon.	Octomeria.	Polystachya.	Xylobium.

the three family ones are Latin, although the central one is originally Greek also.

I adopt this as far as possible for a law through my whole plant nomenclature.

13. Farther: the terminations of the Latin family names will be, for the most part, of the masculine, feminine, and neuter forms, us, a, um, with these following attached conditions.

(I.) Those terminating in 'us,' though often of feminine words, as the central Arbor, will indicate either real masculine strength (quercus, laurus), or conditions of dominant majesty (cedrus), of stubbornness and enduring force (crataegus), or of peasant-like commonalty and hardship (juncus); softened, as it may sometimes happen, into gentleness and beneficence (thymus). The occasional forms in 'er' and 'il' will have similar power (acer, basil).

(II.) Names with the feminine termination 'a,' if they are real names of girls, will always mean flowers that are perfectly pretty and perfectly good, (Lucia, Viola, Margarita, Clarissa). Names terminating in 'a' which are not also accepted names of girls, may sometimes be none the less honourable, (Primula, Campanula,) but for the most part will signify either plants that are only good and worthy in a nursy sort of way, (Salvia,) or that are good without being pretty, (Lavandula,) or pretty without being good, (Kalmia). But no name terminating in 'a' will be attached to a plant that is neither good nor pretty.

(III.) The neuter names terminating in 'um' will always indicate some power either of active or suggestive evil, (Conium, Solanum, Satyrium,) or a relation, more or less definite, to death; but this relation to death may sometimes be noble, or pathetic,—"which to-day is, and to-morrow is cast into the oven,"—Lilium.

But the leading position of the neuters in the plant's double name must be noticed by students unacquainted with Latin, in order to distinguish them from plural genitives, which will always, of course, be the second word (Francesca Fontium, Francesca of the Springs).

14. Names terminating in 'is' and 'e,' if definitely names of women, (Iris, Amaryllis, Alcestis, Daphne,) will always signify flowers of great beauty, and noble historic association. If not definitely names of women, they will yet indicate some speciality of sensitiveness, or association with legend (Berberis, Clematis). No neuters in 'e' will be admitted.

15. Participial terminations (Impatiens), with neuters in 'en' (Cyclamen), will always be descrip-

tive of some special quality or form,—leaving it indeterminate if good or bad, until explained. It will be manifestly impossible to limit either these neuters or the feminines in 'is' to Latin forms; but we shall always know by their termination that they cannot be generic names, if we are strict in forming these last on a given method.

16. How little method there is in our present formation of them, I am myself more and more surprised as I consider. A child is shown a rose, and told that he is to call every flower like that, 'Rosaceous';* he is next shown a lily, and told that he is to call every flower like that, 'Liliaceous';—so far well; but he is next shown a daisy, and is not at all allowed to call every flower like that, 'Daisaceous,' but he must call it, like the fifth order of architecture, 'Composite'; and being next shown a pink, he is not allowed to call other pinks 'Pinkaceous,' but 'Nut-leaved'; and being next shown a pease-blossom, he is not allowed to call other pease-blossoms 'Peasaceous,' but, in a brilliant burst of botanical imagination, he is incited to call it by two names instead of one, 'Butterfly-aceous' from its flower, and 'Pod-aceous' from its seed;—the inconsistency of the terms thus enforced upon him being perfected

* Compare Chapter V., § 7.

in their inaccuracy, for a daisy is not one whit more composite than Queen of the Meadow, or Jura Jacinth;* and 'legumen' is not Latin for a pod, but 'siliqua,'—so that no good scholar could remember Virgil's 'siliqua quassante legumen,' without overthrowing all his Pisan nomenclature.

17. Farther. If we ground our names of the higher orders on the distinctive characters of *form* in plants, these are so many, and so subtle, that we are at once involved in more investigations than a young learner has ever time to follow successfully, and they must be at all times liable to dislocations and rearrangements on the discovery of any new link in the infinitely entangled chain. But if we found our higher nomenclature at once on historic fact, and relative conditions of climate and character, rather than of form, we may at once distribute our flora into unalterable groups, to which we may add at our pleasure, but which will never need disturbance; far less, reconstruction.

18. For instance,—and to begin,—it is an historical fact that for many centuries the English nation believed that the Founder of its religion, spiritually, by the mouth of the King who spake of all herbs, had likened Himself to two flowers, —the Rose of Sharon, and Lily of the Valley.

* 'Jacinthus Jurae,' changed from 'Hyacinthus Comosus.'

The fact of this belief is one of the most important in the history of England,—that is to say, of the mind or heart of England: and it is connected solemnly with the heart of Italy also, by the closing cantos of the Paradiso.

I think it well therefore that our two first generic, or at least commandant, names heading the out-laid and in-laid divisions of plants, should be of the rose and lily, with such meaning in them as may remind us of this fact in the history of human mind.

It is also historical that the personal appearing of this Master of our religion was spoken of by our chief religious teacher in these terms: "The Grace of God, that bringeth salvation, hath appeared unto all men." And it is a constant fact that this 'grace' or 'favour' of God is spoken of as "giving us to eat of the Tree of Life."

19. Now, comparing the botanical facts I have to express, with these historical ones, I find that the rose tribe has been formed among flowers, not in distant and monstrous geologic æras, but in the human epoch;—that its 'grace' or favour has been in all countries so felt as to cause its acceptance everywhere for the most perfect physical type of womanhood;—and that the characteristic fruit of the tribe is so sweet, that it has become

symbolic at once of the subtlest temptation, and the kindest ministry to the earthly passion of the human race. "Comfort me with apples, for I am sick of love."

20. Therefore I shall call the entire order of these flowers 'Charites,' (Graces,) and they will be divided into these five genera, Rosa, Persica, Pomum, Rubra, and Fragaria. Which sequence of names I do not think the young learner will have difficulty in remembering; nor in understanding why I distinguish the central group by the fruit instead of the flower. And if he once clearly master the structure and relations of these five genera, he will have no difficulty in attaching to them, in a satellitic or subordinate manner, such inferior groups as that of the Silverweed, or the Tormentilla; but all he will have to learn by heart and rote, will be these six names; the Greek Master-name, Charites, and the five generic names, in each case belonging to plants, as he will soon find, of extreme personal interest to him.

21. I have used the word 'Order' as the name of our widest groups, in preference to 'Class,' because these widest groups will not always include flowers like each other in form, or equal to each other in vegetative rank; but they will be

'Orders,' literally like those of any religious or chivalric association, having some common link rather intellectual than national,—the Charites, for instance, linked by their kindness,—the Oreiades, by their mountain seclusion, as Sisters of Charity or Monks of the Chartreuse, irrespective of ties of relationship. Then beneath these orders will come, what may be rightly called, either as above in Greek derivation, 'Genera,' or in Latin, 'Gentes,' for which, however, I choose the Latin word, because Genus is disagreeably liable to be confused on the ear with 'genius'; but Gens, never; and also 'nomen gentile' is a clearer and better expression than 'nomen generosum,' and I will not coin the barbarous one, 'genericum.' The name of the Gens, (as 'Lucia,') with an attached epithet, as 'Verna,' will, in most cases, be enough to characterize the individual flower; but if farther subdivision be necessary, the third order will be that of Families, indicated by a 'nomen familiare' added in the third place of nomenclature, as Lucia Verna,—Borealis; and no farther subdivision will ever be admitted. I avoid the word 'species'—originally a bad one, and lately vulgarized beyond endurance—altogether. And varieties belonging to narrow localities, or induced by horticulture, may be named as they please by the people living near the spot, or by the gardener who grows them; but will not be acknowledged by Proserpina. Nevertheless, the arbitrary reduction under Ordines, Gentes, and Familiæ, is always to be remembered as one of massive practical convenience only; and the more subtle arborescence of the infinitely varying structures may be followed, like a human genealogy, as far as we please, afterwards; when once we have got our common plants clearly arranged and intelligibly named.

22. But now we find ourselves in the presence of a new difficulty, the greatest we have to deal with in the whole matter.

Our new nomenclature, to be thoroughly good, must be acceptable to scholars in the five great languages, Greek, Latin, French, Italian, and English; and it must be acceptable by them in teaching the native children of each country. I shall not be satisfied, unless I can feel that the little maids who gather their first violets under the Acropolis rock, may receive for them Æschylean words again with joy. I shall not be content, unless the mothers watching their children at play in the Ceramicus of Paris, under the scarred ruins of her Kings' palace, may yet teach them there to know the flowers which the Maid of Orleans gathered at Domremy. I shall not be satisfied unless every word I ask from the lips of the children of Florence and Rome, may enable them better to praise the flowers that are chosen by the hand of Matilda,* and bloom around the tomb of Vergil.

23. Now in this first example of nomenclature, the Master-name, being *pure* Greek, may easily be accepted by Greek children, remembering that certain also of their own poets, if they did not call the flower a Grace itself, at least thought of it as giving gladness to the Three in their dances.† But for French children the word 'Grâce' has been doubly and trebly corrupted; first, by entirely false theological scholarship, mistaking the 'Favour' or Grace done by God to good men, for the 'Misericordia,' or mercy, shown by Him to bad ones; and so, in practical life, finally substituting 'Grâce' as a word of extreme and mortal prayer, for 'Merci,' and of late using 'Merci' in a totally ridiculous and perverted power, for the giving of thanks, (or refusal of offered good): while the literally derived word 'Charité' has become, in the modern mind, a gift, whether from God or man, only to the wretched, never to the happy: and lastly, 'Grâce' in its physical sense has been perverted, by their

* "Cantando, e scegliendo fior di fiore
Onde era picta tutta la sua via."

Purg., xxviii. 35.

† "*καὶ θεοῖσι τερπνά.*"

social vulgarity, into an idea, whether with respect to form or motion, commending itself rather to the ballet-master than either to the painter or the priest.

For these reasons, the Master name of this family, for my French pupils, must be simply 'Rhodiades,' which will bring, for them, the entire group of names into easily remembered symmetry; and the English form of the same name, Rhodiad, is to be used by English scholars also for all tribes of this group except the five principal ones.

24. Farther, in every gens of plants, one will be chosen as the representative, which, if any, will be that examined and described in the course of this work, if I have opportunity of doing so.

This representative flower will always be a wild one, and of the simplest form which completely expresses the character of the plant; existing divinely and unchangeably from age to age, ungrieved by man's neglect, and inflexible by his power.

And this divine character will be expressed by the epithet 'Sacred,' taking the sense in which we attach it to a dominant and christened majesty, when it belongs to the central type of any forceful order;—'Quercus sacra,' 'Laurus sacra,' etc.,—the word 'Benedicta,' or 'Benedictus,' being used instead, if the plant be too humble to bear, without some discrepancy and unbecomingness, the higher title; as 'Carduus Benedictus,' Holy Thistle.

25. Among the gentes of flowers bearing girls' names, the dominant one will be simply called the Queen, 'Rosa Regina,' 'Rose the Queen' (the English wild rose); 'Clarissa Regina,' 'Clarissa the Queen' (Mountain Pink); 'Lucia Regina,' 'Lucy the Queen' (Spring Gentian), or in simpler English, 'Lucy of Teesdale,' as 'Harry of Monmouth.' The ruling flowers of groups which bear names not yet accepted for names of girls, will be called simply 'Domina,' or shortly 'Donna.' 'Rubra domina' (wild raspberry): the wild strawberry, because of her use in heraldry, will bear a name of her own, exceptional, 'Cora coronalis.'

26. These main points being understood, and concessions made, we may first arrange the greater orders of land plants in a group of twelve, easily remembered, and with very little forcing. There must be *some* forcing always to get things into quite easily tenable form, for Nature always has her ins and outs. But it is curious how fitly and frequently the number of twelve may be used for memoria technica; and in this instance the Greek derivative names fall at once into harmony with the most beautiful parts of Greek mythology, leading on to early Christian tradition.

27. Their series will be, therefore, as follows; the principal subordinate groups being at once placed under each of the great ones. The reasons for occasional appearance of inconsistency will be afterwards explained, and the English and French forms given in each case are the terms which would be used in answering the rapid question, 'Of what order is this flower?' the answer being, It is a 'Cyllenid,' a 'Pleiad,' or a 'Vestal,' as one would answer of a person, he is a Knight of St. John or Monk of St. Benedict; while to the question, of what gens? we answer, a Stella or an Erica, as one would answer for a person, a Stuart or Plantagenet.

I. CHARITES.

Eng. CHARIS. Fr. RHODIADE.

Rosa. Persica. Pomum. Rubra. Fragaria.

II. URANIDES.

Eng. URANID. Fr. URANIDE.

Lucia. Campanula. Convoluta.

III. CYLLENIDES.

Eng. CYLLENID. Fr. NEPHELIDE.

Stella. Francesca. Primula.

IV. OREIADES.

Eng. OREIAD. Fr. OREADE.

Erica. Myrtilla. Aurora.

v. PLEIADES.

Eng. PLEIAD. Fr. PLEIADE.

Silvia. Anemone.

vi. ARTEMIDES.

Eng. ARTEMID. Fr. ARTEMIDE.

Clarissa. Lychnis. Scintilla. Mica.

vii. VESTALES.

Eng. VESTAL. Fr. VESTALE.

Mentha. Melitta. Basil. Salvia. Lavandula. Thymus.

viii. CYTHERIDES.

Eng. CYTHERID. Fr. CYTHERIDE.

Viola. Veronica. Giulietta.

ix. HELIADES.

Eng. ALCESTID. Fr. HELIADE.

Clytia. Margarita. Alcestis. Falconia. Carduus.

x. DELPHIDES.

Eng. DELPHID. Fr. DELPHIDE.

Laurus. Granata. Myrtus.

xi. HESPERIDES.

Eng. HESPERID. Fr. HESPERIDE.

Aurantia. Aegle.

xi. ATHENAIDES.

Eng. ATHENAID Fr. ATHENAIDE.

Olea. Fraxinus.

I will shortly note the changes of name in their twelve orders, and the reasons for them.

I. CHARITES.—The only change made in the nomenclature of this order is the slight one of 'rubra' for 'rubus': partly to express true sisterhood with the other Charites; partly to enforce the idea of redness, as characteristic of the race, both in the lovely purple and russet of their winter leafage, and in the exquisite bloom of scarlet on the stems in strong young shoots. They have every right to be placed among the Charites, first because the raspberry is really a more important fruit in domestic economy than the strawberry; and, secondly, because the wild bramble is often in its wandering sprays even more graceful than the rose; and in blossom and fruit the best autumnal gift that English Nature has appointed for her village children.

II. URANIDES.—Not merely because they are all of the colour of the sky, but also sacred to Urania in their divine purity. 'Convoluta' instead of 'convolvulus,' chiefly for the sake of euphony; but also because Pervinca is to be included in this group.

III. CYLLENIDES.—Named from Mount Cyllene in Arcadia, because the three races included in the order alike delight in rocky ground, and in the cold or moist air of mountain-clouds.

IV. OREIADES.—Described in next chapter.

V. PLEIADES.—From the habit of the flowers belonging to this order to get into bright local clusters. . Silvia, for the wood-sorrel, will I hope be an acceptable change to my girl-readers.

VI. ARTEMIDES.—Dedicate to Artemis for their expression of energy, no less than purity. This character was rightly felt in them by whoever gave the name 'Dianthus' to their leading race; a name which I should have retained if it had not been bad Greek. I wish them, by their name 'Clarissa,' to recall the memory of St. Clare, as 'Francesca' that of St. Francis.* The 'issa,' not without honour to the greatest of our English moral story-tellers, is added for the practical reason, that I think the sound will fasten in the minds of children the essential characteristic of

* The four races of this order are more naturally distinct than botanists have recognized. In Clarissa, the petal is cloven into a fringe at the outer edge; in Lychnis, the petal is terminated in two rounded lobes, and the fringe withdrawn to the top of the limb; in Scintilla, the petal is divided into two *sharp* lobes, without any fringe of the limb; and in Mica, the minute and scarcely visible flowers have simple and far separate petals. The confusion of these four great natural races under the vulgar or accidental botanical names of spittle-plant, shore-plant, sand-plant, etc., has become entirely intolerable by any rational student; but the names 'Scintilla,' substituted for Stellaria, and 'Mica' for the utterly ridiculous and probably untrue Sagina, connect themselves naturally with Lychnis, in expression of the luminous power of the white and sparkling blossoms.

the race, the cutting of the outer edge of the petal as if with scissors.

VII. VESTALES.—I allow this Latin form, because Hestiades would have been confused with Heliades. The order is named 'of the hearth,' from its manifold domestic use, and modest blossoming.

VIII. CYTHERIDES.—Dedicate to Venus, but in all purity and peace of thought. Giuletta, for the coarse, and more than ordinarily false, Polygala.

IX. HELIADES.—The sun-flowers.* In English, Alcestid, in honour to Chaucer and the Daisy.

X. DELPHIDES.—Sacred to Apollo. Granata, changed from Punica, in honour to Granada and the Moors.

XI. HESPERIDES.—Already a name given to the order. Aegle, prettier and more classic than Limonia, includes the idea of brightness in the blossom.

XII. ATHENAIDES.—I take Fraxinus into this group, because the mountain ash, in its hawthorn-scented flower, scarletest of berries, and exquisitely formed and finished leafage, belongs wholly to the floral decoration of our native rocks, and is associated with their human interests, though lightly, not less spiritually, than the olive with the mind of Greece.

* Clytia will include all the true sun-flowers, and Falconia the hawkweeds; but I have not yet completed the analysis of this vast and complex order, so as to determine the limits of Margarita and Alcestis.

28. The remaining groups are in great part natural; but I separate for subsequent study five orders of supreme domestic utility, the Mallows, Currants, Pease,* Cresses, and Cranesbills, from those which, either in fruit or blossom, are for finer pleasure or higher beauty. I think it will be generally interesting for children to learn those five names as an easy lesson, and gradually discover, wondering, the world that they include. I will give their terminology at length, separately.

29. One cannot, in all groups, have all the divisions of equal importance; the Mallows are only placed with the other four for their great value in decoration of cottage gardens in autumn: and their softly healing qualities as a tribe. They will mentally connect the whole useful group with the three great Æsculapiadæ, Cinchona, Coffea, and Camellia.

30. Taking next the water-plants, crowned in the DROSIDÆ, which include the five great families, Juncus, Jacinthus, Amaryllis, Iris, and Lilium, and are masculine in their Greek name because their two first groups, Juncus and Jacinthus, are masculine, I

* The reader must observe that the positions given in this more developed system to any flower do not interfere with arrangements either formerly or hereafter given for memoria technica. The name of the pea, for instance (alata), is to be learned first among the twelve cinqfoils, p. 214, above; then transferred to its botanical place.

gather together the three orders of TRITONIDES, which are notably trefoil; the NAIADES, notably quatrefoil, but for which I keep their present pretty name; and the BATRACHIDES,* notably cinqfoil, for which I keep their present ugly one, only changing it from Latin into Greek.

31. I am not sure of being forgiven so readily for putting the Grasses, Sedges, Mosses, and Lichens together, under the great general head of Demetridæ. But it seems to me the mosses and lichens belong no less definitely to Demeter, in being the first gatherers of earth on rock, and the first coverers of its sterile surface, than the grass which at last prepares it to the foot and to the food of man. And with the mosses I shall take all the especially moss-plants which otherwise are homeless or companionless, Drosera, and the like, and as a connecting link with the flowers belonging to the Dark Kora, the two strange orders of the Ophryds and Agarics.

32. Lastly will come the orders of flowers which may be thought of as belonging for the most part to the Dark Kora of the lower world, —having at least the power of death, if not its terror, given them, together with offices of comfort

* The amphibious habit of this race is to me of more importance than its outlaid structure.

and healing in sleep, or of strengthening, if not too prolonged, action on the nervous power of life. Of these, the first will be the DIONYSIDÆ,—Hedera, Vitis, Liana; then the DRACONIDÆ,—Atropa, Digitalis, Linaria; and, lastly, the MOIRIDÆ,—Conîum, Papaver, Solanum, Arum, and Nerium.

33. As I see this scheme now drawn out, simple as it is, the scope of it seems not only far too great for adequate completion by my own labour, but larger than the time likely to be given to botany by average scholars would enable them intelligently to grasp: and yet it includes, I suppose, not the tenth part of the varieties of plants respecting which, in competitive examination, a student of physical science is now expected to know, or at least assert on hearsay, *something*.

So far as I have influence with the young, myself, I would pray them to be assured that it is better to know the habits of one plant than the names of a thousand; and wiser to be happily familiar with those that grow in the nearest field, than arduously cognisant of all that plume the isles of the Pacific, or illumine the Mountains of the Moon.

Nevertheless, I believe that when once the general form of this system in Proserpina has been well learned, much other knowledge may be easily attached to it, or sheltered under the eaves of it: and in its own development, I believe everything may be included that the student will find useful, or may wisely desire to investigate, of properly European botany. But I am convinced that the best results of his study will be reached by a resolved adherence to extreme simplicity of primal idea, and primal nomenclature.

34. I do not think the need of revisal of our present scientific classification could be more clearly demonstrated than by the fact that laurels and roses are confused, even by Dr. Lindley, in the mind of his feminine readers; the English word laurel, in the index to his first volume of Ladies' Botany, referring them to the cherries, under which the common laurel is placed as 'Prunus Lauro-cerasus,' while the true laurel, 'Laurus nobilis,' must be found in the index of the second volume, under the Latin form 'Laurus.'

This accident, however, illustrates another, and a most important point to be remembered, in all arrangements whether of plants, minerals, or animals. No single classification can possibly be perfect, or anything *like* perfect. It must be, at its best, a ground, or *warp* of arrangement only, through which, or over which, the cross threads of another,—yes, and of many others,—must be woven in our minds. Thus the almond, though in the form and colour of its flower, and method of its fruit, rightly associated with the roses, yet by the richness and sweetness of its kernel must be held mentally connected with all plants that bear nuts. These assuredly must have something in their structure common, justifying their being gathered into a conceived or conceivable group of 'Nuciferæ,' in which the almond, hazel, walnut, cocoa-nut, and such others would be considered as having relationship, at least in their power of secreting a crisp and sweet substance which is not wood, nor bark, nor pulp, nor seed-pabulum reducible to softness by boiling;—but a quite separate substance, for which I do not know that there at present exists any botanical name,—of which, hitherto, I find no general account, and can only myself give so much, on reflection, as that it is crisp and close in texture, and always contains some kind of oil or milk.

35. Again, suppose the arrangement of plants could, with respect to their flowers and fruits, be made approximately complete, they must instantly be broken and reformed by comparison of their stems and leaves. The three *creeping* families of the Charites,—Rosa, Rubra, and Fragaria,—must

then be frankly separated from the elastic Persica and knotty Pomum; of which one wild and lovely species, the hawthorn, is no less notable for the massive accumulation of wood in the stubborn stem of it, than the wild rose for her lovely power of wreathing her garlands at pleasure wherever they are fairest, the stem following them and sustaining, where they will.

36. Thus, as we examine successively each part of any plant, new sisterhoods, and unthought-of fellowships, will be found between the most distant orders; and ravines of unexpected separation open between those otherwise closely allied. Few botanical characters are more definite than the leaf structure illustrated in Plate VI., which has given to one group of the Drosidæ the descriptive name of Ensatæ, (see above, Chapter IX., § 11,) but this conformation would not be wisely permitted to interfere in the least with the arrangement founded on the much more decisive floral aspects of the Iris and Lily. So, in the fifth volume of 'Modern Painters,' the sword-like, or rather rapier-like, leaves of the pine are opposed, for the sake of more vivid realization, to the shield-like leaves of the greater number of inland trees; but it would be absurd to allow this difference any share in botanical arrangement,—else we should find ourselves thrown

into sudden discomfiture by the wide-waving and opening foliage of the palms and ferns.

37. But through all the defeats by which insolent endeavours to sum the orders of Creation must be reproved, and in the midst of the successes by which patient insight will be surprised, the fact of the *confirmation* of species in plants and animals must remain always a miraculous one. What outstretched sign of constant Omnipotence can be more awful, than that the susceptibility to external influences, with the reciprocal power of transformation, in the organs of the plant; and the infinite powers of moral training and mental conception over the nativity of animals, should be so restrained within impassable limits, and by inconceivable laws, that from generation to generation, under all the clouds and revolutions of heaven with its stars, and among all the calamities and convulsions of the Earth with her passions, the numbers and the names of her Kindred may still be counted for her in unfailing truth;—still the fifth sweet leaf unfold for the Rose, and the sixth spring for the Lily; and yet the wolf rave tameless round the folds of the pastoral mountains, and yet the tiger flame through the forests of the night!

CHAPTER XII.

CORA AND KRONOS.

1. OF all the lovely wild plants—and few, mountain-bred, in Britain, are other than lovely,—that fill the clefts and crest the ridges of my Brantwood rock, the dearest to me, by far, are the clusters of whortleberry which divide possession of the lower slopes with the wood hyacinth and pervenche. They are personally and specially dear to me for their association in my mind with the woods of Montanvert; but the plant itself, irrespective of all accidental feeling, is indeed so beautiful in all its ways—so delicately strong in the spring of its leafage, so modestly wonderful in the formation of its fruit, and so pure in choice of its haunts, not capriciously or unfamiliarly, but growing in luxuriance through all the healthiest and sweetest seclusion of mountain territory throughout Europe,—that I think I may without any sharp remonstrance be permitted to express, for this once only, personal feeling in my nomenclature,

calling it in Latin 'Myrtilla Cara,' and in French 'Myrtille Chérie,' but retaining for it in English its simply classic name, 'Blue Whortle.'

2. It is the most common representative of the group of Myrtillæ, which on reference to our classification will be found central between the Ericæ and Auroræ. The distinctions between these three families may be easily remembered, and had better be learned before going farther; but first let us note their fellowship. They are all Oreiades, mountain plants; in specialty, they are all strong in stem, low in stature, and the Ericæ and Auroræ glorious in the flush of their infinitely exulting flowers, ("the rapture of the heath"—above spoken of, p. 96). But all the essential loveliness of the Myrtillæ is in their leaves and fruit: the first always exquisitely finished and grouped like the most precious decorative work of sacred painting; the second, red or purple, like beads of coral or amethyst. Their minute flowers have rarely any general part or power in the colours of mountain ground; but, examined closely, they are one of the chief joys of the traveller's rest among the Alps; and full of exquisiteness unspeakable, in their several bearings and miens of blossom, so to speak. Plate VIII. represents, however feebly, the proud bending back of her head by Myrtilla

VIII.

MYRTILLA REGINA.

Sketched for her gesture only. Isella. 1877.

Regina*: an action as beautiful in *her* as it is terrible in the Kingly Serpent of Egypt.

3. The formal differences between these three families are trenchant and easily remembered. The Ericæ are all quatrefoils, and quatrefoils of the most studied and accomplished symmetry; and they bear no berries, but only dry seeds. The Myrtillæ and Auroræ are both Cinqfoil; but the Myrtillæ are symmetrical in their blossom, and the Auroræ unsymmetrical. Farther, the Myrtillæ are not absolutely determinate in the number of their foils, (this being essentially a characteristic of flowers exposed to much hardship,) and are thus sometimes quatrefoil, in sympathy with the Ericæ. But the Auroræ are strictly cinqfoil. These last are the only European form of a larger group, well named 'Azalea' from the Greek **ἄζα**, dryness, and its adjective *ἀζαλέα*, dry or parched; and *this* name must be kept for the world-wide group, (including under it Rhododendron, but not Kalmia,) because there is an under-meaning in the word Aza, enabling it to be applied to the substance of dry earth, and indicating one of the great functions of the Oreiades, in common with the mosses,—the collection of earth upon rocks.

* 'Arctostaphylos Alpina,' I believe; but scarcely recognize the flower in my botanical books.

4. Neither the Ericæ, as I have just said, nor Auroræ bear useful fruit; and the Ericæ are named from their consequent worthlessness in the eyes of the Greek farmer; they were the plants he 'tore up' for his bed, or signal-fire, his word for them including a farther sense of crushing or bruising into a heap. The Westmoreland shepherds now, alas! burn them remorselessly on the ground, (and a year since had nearly set the copse of Brantwood on fire just above the house). The sense of parched and fruitless existence is given to the heaths, with beautiful application of the context, in our English translation of Jeremiah xvii. 6; but I find the plant there named is, in the Septuagint, Wild Tamarisk; the mountains of Palestine being, I suppose, in that latitude, too low for heath, unless in the Lebanon.

5. But I have drawn the reader's thoughts to this great race of the Orciades at present, because they place for us in the clearest light a question which I have finally to answer before closing the first volume of Proserpina: namely, what is the real difference between the three ranks of Vegetative Humility, and Noblesse—the Herb, the Shrub, and the Tree?

6. Between the herb, which perishes annually, and the plants which construct year after year

an increasing stem, there is, of course, no difficulty of discernment; but between the plants which, like these Oreiades, construct for themselves richest intricacy of supporting stem, yet scarcely rise a fathom's height above the earth they gather and adorn,—between these, and the trees that lift cathedral aisles of colossal shade on Andes and Lebanon,—where is the limit of kind to be truly set?

7. We have the three orders given, as no botanist could, in twelve lines by Milton:—

"Then herbs of every leaf, that sudden flow'r'd,
Op'ning their various colours, and made gay
Her bosom swelling sweet; and, these scarce blown,
Forth flourish'd thick the clust'ring vine, forth crept
The swelling gourd, up stood the corny reed
Embattel'd in her field; and th' *humble shrub,*
And bush with frizzled hair implicit: last
Rose, as in dance, the stately trees, and spread
Their branches hung with copious fruits, or gemm'd
Their blossoms; with high woods the hills were crown'd;
With tufts the valleys and each fountain side;
With borders long the rivers."

Only to learn, and be made to understand, these

twelve lines thoroughly would teach a youth more of true botany than an entire Cyclopædia of modern nomenclature and description: they are, like all Milton's work, perfect in accuracy of epithet, while consummate in concentration. Exquisite in touch, as infinite in breadth, they gather into their unbroken clause of melodious compass the conception at once of the Columbian prairie, the English cornfield, the Syrian vineyard, and the Indian grove. But even Milton has left untold, and for the instant perhaps unthought of, the most solemn difference of rank between the low and lofty trees, not in magnitude only, nor in grace, but in duration.

8. Yet let us pause before passing to this greater subject, to dwell more closely on what he has told us so clearly,—the difference in Grace, namely, between the trees that rise 'as in dance,' and 'the bush with frizzled hair.' For the bush form is essentially one taken by vegetation in some kind of distress; scorched by heat, discouraged by darkness, or bitten by frost; it is the form in which isolated knots of earnest plant life stay the flux of fiery sands, bind the rents of tottering crags, purge the stagnant air of cave or chasm, and fringe with sudden hues of unhoped spring the Arctic edge of retreating desolation.

On the other hand, the trees which, as in sacred

upon a smooth bed of verdure. Between the tropics, the strength and luxury of vegetation give such a development to plants, that the smallest of the dicotyledonous family become shrubs.* It would seem as if the liliaceous plants, mingled with the gramina, assumed the place of the flowers of our meadows. Their form is indeed striking; they dazzle by the variety and splendour of their colours; but, too high above the soil, they disturb that harmonious relation which exists among the plants that compose our meadows and our turf. Nature, in her beneficence, has given the landscape under every zone its peculiar type of beauty.

"After proceeding four hours across the savannahs, we entered into a little wood composed of shrubs and small trees, which is called El Pejual; no doubt because of the great abundance of the 'Pejoa,' (Gaultheria odorata,) a plant with very odoriferous leaves. The steepness of the mountain became less considerable, and we felt an indescribable pleasure in examining the plants of this region. Nowhere, perhaps, can be found collected together in so small a space of ground, productions so beautiful, and so remarkable in regard to the

* I do not see what this can mean. Primroses and cowslips can't become shrubs; nor can violets, nor daisies, nor any other of our pet meadow flowers.

geography of plants. At the height of a thousand toises, the lofty savannahs of the hills terminate in a zone of shrubs, which by their appearance, their tortuous branches, their stiff leaves, and the dimensions and beauty of their purple flowers, remind us of what is called in the Cordilleras of the Andes the vegetation of the *paramos** and the *punas*. We find there the family of the Alpine rhododendrons, the thibaudias, the andromedas, the vacciniums, and those befarias† with resinous leaves, which we have several times compared to the rhododendron of our European Alps.

"Even when nature does not produce the same species in analogous climates, either in the plains of isothermal parallels, or on table-lands the temperature of which resembles that of places nearer the poles, we still remark a striking resemblance of appearance and physiognomy in the vegetation of the most distant countries. This phenomenon is one of the most curious in the history of organic forms. I say the history; for in vain would reason forbid man to form hypotheses on the origin of things: he is not

* 'Deserts.' Punas is not in my Spanish dictionary, and the reference to a former note is wrong in my edition of Humboldt, vol. iii., p. 490.

† "The Alpine rose of equinoctial America," p. 453.

the less tormented with these insoluble problems of the distribution of beings."

15. Insoluble—yes, assuredly, poor little beaten phantasms of palpitating clay that we are—and who asked us to solve it? Even this Humboldt, quiet-hearted and modest watcher of the ways of Heaven, in the real make of him, came at last to be so far puffed up by his vain science in declining years that he must needs write a Kosmos of things in the Universe, forsooth, as if he knew all about them! when he was not able meanwhile, (and does not seem even to have desired the ability,) to put the slightest Kosmos into his own 'Personal Narrative'; but leaves one to gather what one wants out of its wild growth; or rather, to wash or winnow what may be useful out of its débris, without any vestige either of reference or index; and I must look for these fragmentary sketches of heath and grass through chapter after chapter about the races of the Indian, and religion of the Spaniard,—these also of great intrinsic value, but made useless to the general reader by interspersed experiment on the drifts of the wind and the depths of the sea.

16. But one more fragment out of a note (vol. iii., p. 494) I must give, with reference to an order of the Rhododendrons as yet wholly unknown to me.

"The name of vine tree, 'uvas camaronas' (Shrimp grapes?) is given in the Andes to plants of the genus Thibaudia on account of their *large succulent fruit.* Thus the ancient botanists give the name of Bear's vine, 'Uva Ursi,' and vine of Mount Ida, 'Vitis Idea,' to an Arbutus and Myrtillus which belong, like the Thibaudiæ, to the family of the Ericineæ."

Now, though I have one entire bookcase and half of another, and a large cabinet besides, or about fifteen feet square of books on botany beside me here, and a quantity more at Oxford, I have no means whatever, in all the heap, of finding out what a Thibaudia is like. Loudon's Cyclopædia, the only general book I have, tells me only that it will grow well in camellia houses, that its flowers develope at Christmas, and that they are beautifully varied like a fritillary: whereupon I am very anxious to see them, and taste their fruit, and be able to tell my pupils something intelligible of them,—a new order, as it seems to me, among my Oreiades. But for the present I can make no room for them, and must be content, for England and the Alps, with my single class, Myrtilla, including all the fruit-bearing and (more or less) myrtle-leaved kinds; and Azalea for the fruitless flushing of the loftier tribes; taking the special name 'Aurora' for the red and purple ones of Europe, and resigning the already accepted 'Rhodora' to those of the Andes and Himalaya.

17. Of which also, with help of earnest Indian botanists, I hope nevertheless to add some little history to that of our own Oreiades;—but shall set myself on the most familiar of them first, as I partly hinted in taking for the frontispiece of this volume two unchecked shoots of our commonest heath, in their state of full lustre and decline. And now I must go out and see and think—and for the first time in my life—what becomes of all these fallen blossoms, and where my own mountain Cora hides herself in winter; and where her sweet body is laid in its death.

Think of it with me, for a moment before I go. That harvest of amethyst bells, over all Scottish and Irish and Cumberland hill and moorland; what substance is there in it, yearly gathered out of the mountain winds,—stayed there, as if the morning and evening clouds had been caught out of them and woven into flowers; 'Ropes of sea-sand'—but that is child's magic merely, compared to the weaving of the Heath out of the cloud? And once woven, how much of it is for ever worn by the Earth? What weight of that transparent tissue, half crystal and half comb of honey, lies strewn every year dead under the snow?

I must go and look, and can write no more to-day; nor to-morrow neither. I must gather slowly what I see, and remember; and meantime leaving, to be dealt with afterwards, the difficult and quite separate question of the production of *wood*, I will close this first volume of Proserpina with some necessary statements respecting the operations, serviceable to other creatures than themselves, in which the lives of the noblest plants are ended: honourable in this service equally, though evanescent, some,—in the passing of a breeze—or the dying of a day;—and patient some, of storm and time, serene in fruitful sanctity, through all the uncounted ages which Man has polluted with his tears.

CHAPTER XIII.

THE SEED AND HUSK.

1. NOT the least sorrowful, nor least absurd of the confusions brought on us by unscholarly botanists, blundering into foreign languages, when they do not know how to use their own, is that which has followed on their practice of calling the seed-vessels of flowers 'egg-vessels,'* in Latin; thus involving total loss of the power of the good old English word 'husk,' and the good old French one, 'cosse.' For all the treasuries of plants (see Chapter IV., § 17) may be best conceived, and described, generally, as consisting of 'seed' and 'husk,'—for the most part two or more seeds, in a husk composed of two or more parts, as pease in their shell, pips in an orange, or kernels in a walnut; but whatever their number, or the method of their enclosure, let the student keep clear in his mind, for the base of all study of fructification, the broad distinction

* More literally, "persons to whom the care of eggs is entrusted."

between the seed, as one thing, and the husk as another: the seed, essential to the continuance of the plant's race; and the husk, adapted, primarily, to its guard and dissemination; but secondarily, to quite other and far more important functions.

2. For on this distinction follows another practical one of great importance. A seed may serve, and many do mightily serve, for the food of man, when boiled, crushed, or otherwise industriously prepared by man himself, for his mere *sustenance.* But the *husk* of the seed is prepared in many cases for the delight of his eyes, and the pleasure of his palate, by Nature herself, and is then called a 'fruit.'

3. The varieties of structure both in seed and husk, and yet more, the manner in which the one is contained, and distributed by, the other, are infinite; and in some cases the husk is apparently wanting, or takes some unrecognizable form. But in far the plurality of instances the two parts of the plant's treasury are easily distinguishable, and must be separately studied, whatever their apparent closeness of relation, or, (as in all natural things,) the equivocation sometimes taking place between the one and the other. To me, the especially curious point in this matter is that, while I find the most elaborate accounts given by botanists of

the stages of growth in each of these parts of the treasury, they never say of what use the guardian is to the guarded part, irrespective of its service to man. The mechanical action of the husk in containing and scattering the seeds, they indeed often notice and insist on; but they do not tell us of what, if any, nutritious or fostering use the rind is to a chesnut, or an orange's pulp to its pips, or a peach's juice to its stone.

4. Putting aside this deeper question for the moment, let us make sure we understand well, and define safely, the separate parts themselves. A seed consists essentially of a store, or sack, containing substance to nourish a germ of life, which is surrounded by such substance, and in the process of growth is first fed by it. The germ of life itself rises into two portions, and not more than two, in the seeds of two-leaved plants; but this symmetrical dualism must not be allowed to confuse the student's conception, of the *three* organically separate parts,—the tough skin of a bean, for instance; the softer contents of it which we boil to eat; and the small germ from which the root springs when it is sown. A bean is the best type of the whole structure. An almond out of its shell, a peach-kernel, and an apple-pip are also clear and perfect, though varied types.

5. The husk, or seed-vessel, is seen in perfect simplicity of type in the pod of a bean, or the globe of a poppy. There are, I believe, flowers in which it is absent or imperfect; and when it contains only one seed, it may be so small and closely united with the seed it contains, that both will be naturally thought of as one thing only. Thus, in a dandelion, the little brown grains, which may be blown away, each with its silken parachute, are every one of them a complete husk and seed together. But the majority of instances (and those of plants the most serviceable to man) in which the seed-vessel has entirely a separate structure and mechanical power, justify us in giving it the normal term 'husk,' as the most widely applicable and intelligible.

6. The change of green, hard, and tasteless vegetable substance into beautifully coloured, soft, and delicious substance, which produces what we call a fruit, is, in most cases, of the husk only; in others, of the part of the stalk which immediately sustains the seed; and in a very few instances, not properly a change, but a distinct formation, of fruity substance between the husk and seed. Normally, however, the husk, like the seed, consists always of three parts; it has an outer skin, a central substance of peculiar nature, and an inner

skin, which holds the seed. The main difficulty, in describing or thinking of the completely ripened product of any plant, is to discern clearly which is the inner skin of the husk, and which the outer skin of the seed. The peach is in this respect the best general type,—the woolly skin being the outer one of the husk; the part we eat, the central substance of the husk; and the hard shell of the stone, the inner skin of the husk. The bitter kernel within is the seed.

7. In this case, and in the plum and cherry, the two parts under present examination—husk and seed—separate naturally; the fruity part, which is the body of the husk, adhering firmly to the shell, which is its inner coat. But in the walnut and almond, the two outer parts of the husk separate from the interior one, which becomes an apparently independent 'shell.' So that when first I approached this subject I divided the general structure of a treasury into *three* parts—husk, shell, and kernel; and this division, when we once have mastered the main one, will be often useful. But at first let the student keep steadily to his conception of the two constant parts, husk and seed, reserving the idea of shells and kernels for one group of plants only.

8. It will not be always without difficulty that

he maintains the distinction, when the tree pretends to have changed it. Thus, in the chesnut, the inner coat of the husk becomes brown, adheres to the seed, and seems part of it; and we naturally call only the thick, green, prickly coat, the husk. But this is only one of the deceiving tricks of Nature, to compel our attention more closely. The real place of separation, to *her* mind, is between the mahogany coloured shell and the nut itself, and that more or less silky and flossy coating within the brown shell is the true lining of the entire 'husk.' The paler brown skin, following the rugosities of the nut, is the true sack or skin of the seed. Similarly in the walnut and almond.

9. But, in the apple, two new tricks are played us. First, in the brown skin of the ripe pip, we might imagine we saw the part correspondent to the mahogany skin of the chesnut, and therefore the inner coat of the husk. But it is not so. The brown skin of the pips belongs to them properly, and is all their own. It is the true skin or sack of the seed. The inner coat of the husk is the smooth, white, scaly part of the core that holds them.

Then,—for trick number two. We should as naturally imagine the skin of the apple, which we peel off, to be correspondent to the skin of the peach; and therefore, to be the outer part of the husk. But not at all. The outer part of the husk in the apple is melted away into the fruity mass of it, and the red skin outside is the skin of its *stalk*, not of its seed-vessel at all!

10. I say 'of its stalk,'—that is to say, of the part of the stalk immediately sustaining the seed, commonly called the torus, and expanding into the calyx. In the apple, this torus incorporates itself with the husk completely; then refines its own external skin, and colours *that* variously and beautifully, like the true skin of the husk in the peach, while the withered leaves of the calyx remain in the 'eye' of the apple.

But in the 'hip' of the rose, the incorporation with the husk of the seed does not take place. The torus, or,—as in this flower from its peculiar form it is called,—the tube of the calyx, alone forms the frutescent part of the hip; and the complete seeds, husk and all, (the firm triangular husk enclosing an almond-shaped kernel,) are grouped closely in its interior cavity, while the calyx remains on the top in a large and scarcely withering star. In the nut, the calyx remains green and beautiful, forming what we call the husk of a filbert; and again we find Nature amusing herself by trying to make us think that this strict envelope, almost closing over the single seed, is the same thing to the nut that its green shell is to a walnut!

11. With still more capricious masquing, she varies and hides the structure of her 'berries.'

The strawberry is a hip turned inside-out, the frutescent receptacle changed into a scarlet ball, or cone, of crystalline and delicious coral, in the outside of which the separate seeds, husk and all, are imbedded. In the raspberry and blackberry, the interior mound remains sapless; and the rubied translucency of dulcet substance is formed round each separate seed, *upon* its husk; not a part of the husk, but now an entirely independent and added portion of the plant's bodily form.

12. What is thus done for each seed, on the *out*side of the receptacle, in the raspberry, is done for each seed, *in*side the calyx, in a pomegranate; which is a hip in which the seeds have become surrounded with a radiant juice, richer than claret wine; while the seed itself, within the generous jewel, is succulent also, and spoken of by Tournefort as a "baie succulente." The tube of the calyx, brown-russet like a large hip, externally, is yet otherwise divided, and separated wholly from the cinque-foiled, and

cinque-celled rose, both in number of petal and division of treasuries; the calyx has eight points, and nine cells.

13. Lastly, in the orange, the fount of fragrant juice is interposed between the seed and the husk. It is wholly independent of both; the Aurantine rind, with its white lining and divided compartments, is the true husk: the orange pips are the true seeds; and the eatable part of the fruit is formed between them, in clusters of delicate little flasks, as if a fairy's store of scented wine had been laid up by her in the hollow of a chesnut shell, between the nut and rind; and then the green changed to gold.

14. I have said '*lastly*'—of the orange, for fear of the reader's weariness only; not as having yet represented, far less exhausted, the variety of frutescent form. But these are the most important types of it; and before I can explain the relation between these, and another, too often confounded with them—the *granular* form of the seed of grasses,—I must give some account of what, to man, is far more important than the form—the gift to him in fruit-food; and trial, in fruit-temptation.

CHAPTER XIV.

THE FRUIT GIFT.

1. IN the course of the preceding chapter, I hope that the reader has obtained, or may by a little patience both obtain and secure, the idea of a great natural Ordinance, which, in the protection given to the part of plants necessary to prolong their race, provides, for happier living creatures, food delightful to their taste, and forms either amusing or beautiful to their eyes. Whether in receptacle, calyx, or true husk,—in the cup of the acorn, the fringe of the filbert, the down of the apricot, or bloom of the plum, the powers of Nature consult quite other ends than the mere continuance of oaks and plum trees on the earth; and must be regarded always with gratitude more deep than wonder, when they are indeed seen with human eyes and human intellect.

2. But in one family of plants, the *contents* also of the seed, not the envelope of it merely, are prepared for the support of the higher animal

life: and their grain, filled with the substance which, for universally understood name, may best keep the Latin one of Farina,—becoming in French, 'Farine,' and in English, 'Flour,'—both in the perfectly nourishing elements of it, and its easy and abundant multiplicability, becomes the primal treasure of human economy.

3. It has been the practice of botanists of all nations to consider the seeds of the grasses together with those of roses and pease, as if all could be described on the same principles, and with the same nomenclature of parts. But the grain of corn is a quite distinct thing from the seed of pease. In *it*, the husk and the seed envelope have become inextricably one. All the exocarps, endocarps, epicarps, mesocarps, shells, husks, sacks, and skins, are woven at once together into the brown bran; and inside of that, a new substance is collected for us, which is not what we boil in pease, or poach in eggs, or munch in nuts, or grind in coffee;—but a thing which, mixed with water and then baked, has given to all the nations of the world their prime word for food, in thought and prayer,—Bread; their prime conception of the man's and woman's labour in preparing it—("whoso putteth hand to the *plough*"—two women shall be grinding at the

mill)—their prime notion of the means of cooking by fire—("which to-day is, and to-morrow is cast into the *oven*"), and their prime notion of culinary office—the "chief *baker*," cook, or pastry-cook,—(compare Bedreddin Hassan in the Arabian Nights): and, finally, to modern civilization, the Saxon word 'lady,' with whatever it imports.

4. It has also been the practice of botanists to confuse all the ripened products of plants under the general term 'fruit.' But the essential and separate fruit-gift is of two substances, quite distinct from flour, namely, oil and wine, under the last term including for the moment all kinds of juice which will produce alcohol by fermentation. Of these, oil may be produced either in the kernels of nuts, as in almonds, or in the substance of berries, as in the olive, date, and coffee-berry. But the sweet juice which will become medicinal in wine, can only be developed in the husk, or in the receptacle.

5. The office of the Chief Butler, as opposed to that of the Chief Baker, and the office of the Good Samaritan, pouring in oil and wine, refer both to the total fruit-gift in both kinds: but in the study of plants, we must primarily separate our notion of their gifts to men into the three elements, flour, oil, and wine; and have instantly

and always intelligible names for them in Latin, French, and English.

And I think it best not to confuse our ideas of pure vegetable substance with the possible process of fermentation:—so that rather than 'wine,' for a constant specific term, I will take 'Nectar,'—this term more rightly including the juices of the peach, nectarine, and plum, as well as those of the grape, currant, and apple.

Our three separate substances will then be easily named in all three languages:

Farina.	Oleum.	Nectar.
Farine.	Huile.	Nectare.
Flour.	Oil.	Nectar.

There is this farther advantage in keeping the third common term, that it leaves us the words Succus, Jus, Juice, for other liquid products of plants, watery, milky, sugary, or resinous,—often indeed important to man, but often also without either agreeable flavour or nutritious power; and it is therefore to be observed with care that we may use the word 'juice,' of a liquid produced by any part of a plant, but 'nectar,' only of the juices produced in its fruit.

6. But the good and pleasure of fruit is not in the juice only;—in some kinds, and those not the least valuable, (as the date,) it is not in the

juice at all. We still stand absolutely in want of a word to express the more or less firm *substance* of fruit, as distinguished from all other products of a plant. And with the usual ill-luck—(I advisedly think of it as demoniacal misfortune)—of botanical science, no other name has been yet used for such substance than the entirely false and ugly one of 'Flesh,'—Fr., 'Chair,' with its still more painful derivation 'Charnu,' and in England the monstrous scientific term, 'Sarco-carp.'

But, under the housewifery of Proserpina, since we are to call the juice of fruit, Nectar, its substance will be as naturally and easily called Ambrosia; and I have no doubt that this, with the other names defined in this chapter, will not only be found practically more convenient than the phrases in common use, but will more securely fix in the student's mind a true conception of the essential differences in substance, which, ultimately, depend wholly on their pleasantness to human perception, and offices for human good; and not at all on any otherwise explicable structure or faculty. It is of no use to determine, by microscope or retort, that cinnamon is made of cells with so many walls, or grape-juice of molecules with so many sides;—we are just as far as ever from understanding why these particular interstices should be aromatic, and these special parallelopipeds exhilarating, as we were in the savagely unscientific days when we could only see with our eyes, and smell with our noses. But to call each of these separate substances by a name rightly belonging to it through all the past variations of the language of educated man, will probably enable us often to discern powers in the thing itself, of affecting the human body and mind, which are indeed qualities infinitely more its *own*, than any which can possibly be extracted by the point of a knife, or brayed out with a mortar and pestle.

7. Thus, to take merely instance in the three main elements of which we have just determined the names,—flour, oil, and ambrosia;—the differences in the kinds of pleasure which the tongue received from the powderiness of oat-cake, or a well-boiled potato—(in the days when oat-cake and potatoes were!)—from the glossily-softened crispness of a well-made salad, and from the cool and fragrant amber of an apricot, are indeed distinctions between the essential virtues of things which were made to be *tasted*, much more than to be eaten; and in their various methods of ministry to, and temptation of, human appetites,

have their part in the history, not of elements merely, but of souls; and of the soul-virtues, which from the beginning of the world have bade the barrel of meal not waste, nor the cruse of oil fail; and have planted, by waters of comfort, the fruits which are for the healing of nations.

8. And, again, therefore, I must repeat, with insistence, the claim I have made for the limitation of language to the use made of it by educated men. The word 'carp' could never have multiplied itself into the absurdities of endo-carps and epi-carps, but in the mouths of men who scarcely ever read it in its original letters, and therefore never recognized it as meaning precisely the same thing as 'fructus,' which word, being a little more familiar with, they would have scarcely abused to the same extent; they would not have called a walnut shell an intra-fruct—or a grape skin an extra-fruct; but again, because, though they are accustomed to the English 'fructify,' 'frugivorous'—and 'usufruct,' they are unaccustomed to the Latin 'fruor,' and unconscious therefore that the derivative 'fructus' must always, in right use, mean an *enjoyed* thing, they generalize every mature vegetable product under the term; and we find Dr. Gray coolly telling us that there is no fruit so "likely to be mistaken for a

seed," as a grain of corn! a grain, whether of corn, or any other grass, being precisely the vegetable structure to which frutescent change is for ever forbidden! and to which the word *seed* is primarily and perfectly applicable!—the thing to be *sown*, not grafted.

9. But to mark this total incapability of frutescent change, and connect the form of the seed more definitely with its dusty treasure, it is better to reserve, when we are speaking with precision, the term 'grain' for the seeds of the grasses: the difficulty is greater in French, than in English: because they have no monosyllabic word for the constantly granular 'seed'; but for us the terms are all simple, and already in right use, only not quite clearly enough understood; and there remains only one real difficulty now in our system of nomenclature, that having taken the word 'husk' for the seed-vessel, we are left without a general word for the true fringe of a filbert, or the chaff of a grass. I don't know whether the French 'frange' could be used by them in this sense, if we took it in English botany. But for the present, we can manage well enough without it, one general term, 'chaff,' serving for all the grasses, 'cup' for acorns, and 'fringe' for nuts.

10. But I call this a *real* difficulty, because I

suppose, among the myriads of plants of which I know nothing, there may be forms of the envelope of fruits or seeds which may, for comfort of speech, require some common generic name. One *un*real difficulty, or shadow of difficulty, remains in our having no entirely comprehensive name for seed and seed-vessel together than that the botanists now use, 'fruit.' But practically, even now, people feel that they can't gather figs of thistles, and never speak of the fructification of a thistle, or of the fruit of a dandelion. And, re-assembling now, in one view, the words we have determined on, they will be found enough for all practical service, and in such service always accurate, and, usually, suggestive. I repeat them in brief order, with such farther explanation as they need.

11. All ripe products of the life of flowers consist essentially of the Seed and Husk,—these being, in certain cases, sustained, surrounded, or provided with means of motion, by other parts of the plant; or by developments of their own form which require in each case distinct names. Thus the white cushion of the dandelion to which its brown seeds are attached, and the personal parachutes which belong to each, must be separately described for that species of plants; it is the little

brown thing they sustain and carry away on the wind, which must be examined as the essential product of the floret;—the 'seed and husk.'

12. Every seed has a husk, holding either that seed alone, or other seeds with it.

Every perfect seed consists of an embryo, and the substance which first nourishes that embryo; the whole enclosed in a sack or other sufficient envelope. Three essential parts altogether.

Every perfect husk, vulgarly pericarp, or 'round-fruit,'—(as periwig, 'round-wig,')—consists of a shell, (vulgarly endocarp), rind, (vulgarly mesocarp), and skin, (vulgarly epicarp); three essential parts altogether. But one or more of these parts may be effaced, or confused with another; and in the seeds of grasses they all concentrate themselves into bran.

13. When a husk consists of two or more parts, each of which has a separate shaft and volute, uniting in the pillar and volute of the flower, each separate piece of the husk is called a 'carpel.' The name was first given by De Candolle, and must be retained. But it continually happens that a simple husk divides into two parts corresponding to the two leaves of the embryo, as in the peach, or symmetrically holding alternate seeds, as in the pea. The beautiful drawing of the pea-shell with

its seeds, in Rousseau's botany, is the only one I have seen which rightly shows and expresses this arrangement.

14. A Fruit, is either the husk, receptacle, petal, or other part of a flower *external to the seed*, in which chemical changes have taken place, fitting it for the most part to become pleasant and healthful food for man, or other living animals; but in some cases making it bitter or poisonous to them, and the enjoyment of it depraved or deadly. But, as far as we know, it is without any definite office to the seed it contains; and the change takes place entirely to fit the plant to the service of animals.* In its perfection, the Fruit Gift is limited to a temperate zone, of which the polar limit is marked by the strawberry, and the equatorial by the orange. The more arctic regions produce even the smallest kinds of fruit with difficulty; and the more equatorial, in coarse, oleaginous, or over-luscious masses.

15. All the most perfect fruits are developed

* A most singular sign of this function is given in the chemistry of the changes, according to a French botanist, to whose carefully and richly illustrated volume I shall in future often refer my readers, "Vers l'époque de la maturité, les fruits *exhalent de l'acide carbonique*. Ils ne presentent plus dès lors aucun dégagement d'oxygène pendant le jour, et *respirent, pour ainsi dire, à la façon des animaux*."—(Figuier, 'Histoire des Plantes,' p. 182. 8vo. Paris. Hachette, 1874.)

from exquisite forms either of foliage or flower. The vine leaf, in its generally decorative power, is the most important, both in life and in art, of all that shade the habitations of men. The olive leaf is, without any rival, the most beautiful of the leaves of timber trees; and its blossom, though minute, of extreme beauty. The apple is essentially the fruit of the rose, and the peach of her only rival in her own colour. The cherry and orange blossom are the two types of floral snow.

16. And, lastly, let my readers be assured, the economy of blossom and fruit, with the distribution of water, will be found hereafter the most accurate test of wise national government.

For example of the action of a national government, rightly so called, in these matters, I refer the student to the Mariegolas of Venice, translated in Fors Clavigera; and I close this chapter, and this first volume of Proserpina, not without pride, in the words I wrote on this same matter eighteen years ago. "So far as the labourer's immediate profit is concerned, it matters not an iron filing whether I employ him in growing a peach, or in forging a bombshell. But the difference to him is final, whether, when his child is ill, I walk into his cottage, and give it the peach,—or drop the shell down his chimney, and blow his roof off."

INDEX I.

DESCRIPTIVE NOMENCLATURE.

PLANTS in perfect form are said, at page 29, to consist of four principal parts: root, stem, leaf, and flower. (Compare Chapter V., § 2.) The reader may have been surprised at the omission of the fruit from this list. But a plant which has borne fruit is no longer of 'perfect' form. Its flower is dead. And, observe, it is further said, at page 73, (and compare Chapter III., § 2,) that the use of the fruit is to produce the flower: not of the flower to produce the fruit. Therefore, the plant in perfect blossom, is itself perfect. Nevertheless, the formation of the fruit, practically, is included in the flower, and so spoken of in the fifteenth line of the same page.

Each of these four main parts of a plant consist normally of a certain series of minor parts, to which it is well to attach easily remembered names. In this section of my index I will not admit the confusion of idea involved by alphabetical arrangement of these names, but will sacrifice facility of reference to clearness of explanation, and taking the four great parts of the plant in succession, I will give the list of the minor and constituent parts, with their names as determined in Proserpina, and reference to the pages where the reasons for such determination are given, endeavouring to supply, at the same time, any deficiencies which I find in the body of the text.

I. The Root.

The nomenclature of Roots will not be extended, in Proserpina, beyond the five simple terms here given: though the ordinary botanical ones—corm, bulb, tuber, etc.—will be severally explained in connection with the plants which they specially characterize.

II. THE STEM.

III. THE LEAF.

The nomenclature of the leaf consists, in botanical books, of little more than barbarous, and, for the general reader, totally useless attempts to describe their forms in Latin. But their forms are infinite and indescribable except by the pencil. I will give central types of form in the next volume of Proserpina; which, so that the reader sees and remembers, he may *call* anything he likes. But it is necessary that names should be assigned to certain classes of leaves which are essentially different from each other in character and tissue, not merely in form. Of these the two main divisions have been already given: but I will now add the less important ones which yet require distinct names.

It ought to have been noticed that the character of serration, within reserved limits, is essential to an Apolline leaf, and absolutely refused by an Arethusan one.

III. DRYAD.—Of the ordinary leaf tissue, neither manifestly strong, nor admirably tender, but serviceably consistent, which we find generally to be the substance of

IV. THE FLOWER.

These being all the essential parts of the flower itself, other forms and substances are developed in the seed as it ripens, which, I believe, may most conveniently be arranged in a separate section, though not logically to be considered as separable from the flower, but only as mature states of certain parts of it.

V. THE SEED.

I must once more desire the reader to take notice that, under the four sections already defined, the morphology of the plant is to be considered as complete, and that we are now only to examine and name, farther, its *product;* and that not so much as the germ of its own future descendant flower, but as a separate substance which it is appointed to form, partly to its own detriment, for the sake of higher creatures. This product consists essentially of two parts: the Seed and its Husk.

Proserpina. I delay any extended description of these until we have examined the structure of wood itself more closely; this intricate and difficult task having been remitted (p. 195) to the days of coming spring; and I am well pleased that my younger readers should at first be vexed with no more names to be learned than those of the vegetable productions with which they are most pleasantly acquainted: but for older ones, I think it well, before closing the present volume, to indicate, with warning, some of the obscurities, and probable fallacies, with which this vanity of science encumbers the chemistry, no less than the morphology, of plants.

Looking back to one of the first books in which our new knowledge of organic chemistry began to be displayed, thirty years ago, I find that even at that period the organic elements which the cuisine of the laboratory had already detected in simple Indigo, were the following:—

Isatine,	Chlorindine,
Bromisatine,	Chlorindoptene,
Bibromisatine;	Chlorindatmit;
Chlorisatine,	Chloranile,
Bichlorisatine;	Chloranilam, and,
Chlorisatyde,	Chloranilammon.
Bichlorisatyde;	

And yet, with all this practical skill in decoction, and accumulative industry in observation and nomenclature, so far are our scientific men from arriving, by any decoctive process of their own knowledge, at general results useful to ordinary human creatures, that when I wish now to separate, for young scholars, in first massive arrangement of vegetable productions, the Substances of Plants from their Essences; that is to say, the weighable and measurable body of the plant from its practically immeasurable, if not imponderable, spirit, I find in my three volumes of close-printed chemistry, no information whatever respecting the quality of volatility in matter, except this one sentence:—

"The disposition of various substances to yield vapour is very different: and the difference depends doubtless on the relative power of cohesion with which they are endowed." *

Even in this not extremely pregnant, though extremely cautious, sentence, two conditions of matter are confused, no notice being taken of the difference in manner of dissolution between a vitally fragrant and a mortally putrid substance.

It is still more curious that when I look for

* 'Elements of Chemistry,' p. 44. By Edward Turner; edited by Justus Liebig and William Gregory. Taylor and Walton, 1840.

more definite instruction on such points to the higher ranks of botanists, I find in the index to Dr. Lindley's 'Introduction to Botany'—seven hundred pages of close print—not one of the four words 'Volatile,' 'Essence,' 'Scent,' or 'Perfume.' I examine the index to Gray's 'Structural and Systematic Botany,' with precisely the same success. I next consult Professors Balfour and Grindon, and am met by the same dignified silence. Finally, I think over the possible chances in French, and try in Figuier's indices to the 'Histoire des Plantes' for 'Odeur'—no such word! 'Parfum'—no such word. 'Essence'—no such word. 'Encens'—no such word. I try at last 'Pois de Senteur,' at a venture, and am referred to a page which describes their going to sleep.

Left thus to my own resources, I must be content for the present to bring the subject at least under safe laws of nomenclature. It is possible that modern chemistry may be entirely right in alleging the absolute identity of substances such as albumen, or fibrine, whether they occur in the animal or vegetable economies. But I do not choose to assume this identity in my nomenclature. It may, perhaps, be very fine and very instructive to inform the pupils preparing for competitive examination that the main element of Milk is Milkine, and of Cheese, Cheesine. But for the practical purposes of life, all that I think it necessary for the pupil to know is that in order to get either milk or cheese, he must address himself to a Cow, and not to a Pump; and that what a chemist can produce for him out of dandelions or cocoanuts, however milky or cheesy it may look, may more safely be called by some name of its own.

This distinctness of language becomes every day more desirable, in the face of the refinements of chemical art which now enable the ingenious confectioner to meet the demands of an unscientific person for (suppose,) a lemon drop, with a mixture of nitric acid, sulphur, and stewed bones. It is better, whatever the chemical identity of the products may be, that each should receive a distinctive epithet, and be asked for and supplied, in vulgar English, and vulgar probity, either as essence of lemons, or skeletons.

I intend, therefore,—and believe that the practice will be found both wise and convenient,—to separate in all my works on natural history the terms used for vegetable products from those used for animal or mineral ones, whatever may be their chemical identity, or resemblance in aspect. I do not mean to talk of fat in seeds, nor of flour in eggs, nor of milk in rocks. Pace my prelatical friends, I mean to use the word 'Alb' for vegetable albumen; and although I cannot without pedantry avoid using sometimes the word 'milky' of the white juices of plants, I must beg the reader to remain unaffected in his conviction that there is a vital difference between liquids that coagulate into butter, or congeal into India-rubber. Oil, when used simply, will always mean a vegetable product: and when I have occasion to speak of petroleum, tallow, or blubber, I shall generally call these substances by their right names.

There are also a certain number of vegetable materials more or less prepared, secreted, or digested for us by animals, such as wax, honey, silk, and cochineal. The properties of these require more complex definitions, but they have all very intelligible and well-established names. 'Tea' must be a general term for an extract of any plant in boiling water: though when standing alone the word will take its accepted Chinese meaning: and essence, the general term for the condensed dew of a vegetable vapour, which is with grace and fitness called the 'being' of a plant, because its properties are almost always characteristic of the species; and it is not, like leaf tissue or wood fibre, approximately the same material in different shapes; but a separate element in each family of flowers, of a mysterious, delightful, or dangerous influence, logically inexplicable, chemically inconstructible, and wholly, in dignity of nature, above all modes and faculties of form.

INDEX II.

TO THE PLANTS SPOKEN OF IN THIS VOLUME, UNDER THEIR ENGLISH NAMES ACCEPTED BY PROSERPINA.

INDEX III.

TO THE PLANTS SPOKEN OF IN THIS VOLUME, UNDER THEIR LATIN OR GREEK NAMES, ACCEPTED BY PROSERPINA.

www.ingramcontent.com/pod-product-compliance
Lightning Source LLC
Chambersburg PA
CBHW021218270326
41929CB00010B/1178

9783337111601